DATE			

BAKER & TAYLOR

Minds for the Making

Minds for the Making

The Role of Science in
American Education, 1750–1990

SCOTT L. MONTGOMERY

THE GUILFORD PRESS
New York   London

To Shirley, Marilyn, and Kyle
to whom everything returns

© 1994 The Guilford Press
A Division of Guilford Publications, Inc.
72 Spring Street, New York, NY 10012

Printed in the United States of America

This book is printed on acid-free paper.

Last digit is print number: 9 8 7 6 5 4 3 2 1

Library of Congress Cataloging in Publication Data
Montgomery, Scott L.
    Minds for the making / Scott L. Montgomery
      p.   cm.
    Includes bibliographical references and index.
    ISBN 0-89862-189-5.—ISBN 0-89862-188-7 (pbk.)
    1. Science—Study and teaching—United States—History. 2. Science and state—United States—History. 3. Education—Social aspects—United States—History. I. Title.
Q183.3.A1M66   1994   93-38389
507'.0973—dc20   CIP

Preface

This book, like many others no doubt, began as a kind of accident. It was originally a brief essay on debates over science in the curriculum during the past 50 years. I had been struck how recent controversies in the university seemed to ignore the sciences almost entirely. Intrigued about whether such indifference to science was my own false perception or a radical break from past history, I wanted to look into the matter a bit further. I knew enough to understand that science had always been at the forefront of curriculum debates in the past. What had happened to change this, in image or reality?

The essay I wrote was not particularly satisfactory, neither to me nor to the editor for whom I had proposed it. On the basis of her criticisms, I began to pursue the subject in more detail, looking back through decades, following various trains of possible thought. But I soon found that bringing "science" and "education" together in any kind of interpretive frame, especially a historical one, dragged forward a host of other realities deeply involved in the practice of institutional schooling, realities that have long sealed these two ideas tightly together and that include such things as school structure, testing, psychologies of learning, curriculum theory, ideas of the student, professional and popular images of science, and, of course, to close the circle, history itself. Very soon, my original project got out of hand.

This book, therefore, should be considered a type of cultural criticism. Any history that deals with education is a history about culture and culture making, of course. Inasmuch as it examines learning, it must also bring into the open many other themes that have always stood in relational proximity to this—themes, for example, of identity, nationhood, scholarship, gender and ethnicity, the struggle to give ideals for social conversion a legitimate, institutional form. It has been my aim to try and bring most of these themes to the surface in their relation to both science and the curriculum. Indeed, it has been the projection of such themes onto the images and into the realities of these two aspects of culture that have kept them both such active agents of change and of constancy.

In my effort I have tried to draw on a number of different types of sources. Most helpful and useful among them all, however, was the impressive and far-ranging work of Lawrence Cremin, whose three-volume study and analysis of American education, from the colonial period to the present, should perhaps itself comprise a required course in any university curriculum. As it is, an odd but strategic omission—never discussed—can be found nearly everywhere in our system of higher learning, by which I mean the lack of (required) courses on the university itself. Whatever one's position might be with regard to the "Western" tradition, there remains something striking in the fact that American students are not urged to study the very institution in which they reside, in which knowledge itself has been retained and invested. But of this and other topics, more later.

In most cases, I have tried to provide some measure of context by bringing in developments and concepts related to the larger evolution of American education and American science, and how, over and again, images derived from one profoundly influenced the other. Thus, if the narrative appears to range widely at times, over variable topics and events, this is to suggest the broader outlines of what is essential; the complexity of ideas, ambitions, and results that all played a part in bringing these two areas to their present state.

A work such as this could not have been conceived or completed without debt to others. My thanks go first to Les Levidow and Silvia Federici, whose criticisms of my earlier essay led directly to the writing of this book. To Peter Wissoker of The Guilford Press I express my gratitude for his encouragement and support. And lastly, my unbounded appreciation goes out to my wife

Marilyn, who carefully read successive drafts of the manuscript while suffering its nuptial presence in our lives; to my friend Jeff Eaton who also provided valuable criticism on many points; and finally to Dominic DiBernardi, lifelong companion and intellectual confidant, without whom this and many other things would never have seen print.

Contents

Subject-matter is but spiritual food, possible nutritive material. It cannot digest itself. . . . The source of whatever is dead, mechanical, and formal in schools is found precisely in the subordination of the life and experience of the child to the curriculum. It is because of this fact that "study" has become a synonym for what is irksome, and a lesson identical with task.

—JOHN DEWEY (1902)

Introduction:
"The Field at Our Door"

The role of science in the history of debates surrounding the curriculum in the United States has touched all levels of and concepts about schooling. It has involved the development of science education as well as the concept of a "scientific" education, the training of a powerful elite as well as mass learning. As such, it has drawn upon a wide spectrum of competing interests and ideas. It has helped guide the framing of concepts as complex and influential as "progress," "human nature," "the national welfare," and "learning" itself. It has been a stage on which many players have entered, spoken their parts, exited, and sometimes returned (not always in the guise of farce). But above all, it is something from which a great deal can be learned: In large part any current situation can be related to this long history and cannot in any sense be divorced from what it reveals about the larger place of science and education in the American consciousness. More than a few thinkers, both on the left and right, have correlated the progress of schooling with that of nationalism, noting that education generally seems one of the most ardent and effective champions of the state. But history reveals this connection to be an enormous oversimplification, woefully expedient, and particularly so in the case of science. Here, in fact, "nationalism" itself can be revealed as a collision of many conflicting interests, myths, visions, and hopes, all of which at some point took the "scientific" as legitimating dais. No committed history could be so reductive. It must instead open both inward and outward, toward the past and its continued momentum into the future.

There has probably never been a time in American history during which ardent debate about education and its reform was absent. The curriculum of colleges and universities, in particular, has been a subject of more study, argument, progressivism, and conservative resistance than any discipline within its purview. This, no doubt, is due to the confused but deep-seated American faith in the power of education to be all things and to serve many ends, be they moral, psychological, economic, political, spiritual, or intellectual. As Richard Hofstadter (1963, p. 301) once put it, somewhat wryly, "The history of American educational reformers often seems to be the history of men fighting against an uncongenial environment. . . . That this literature should have been one of complaint is not in itself surprising, for complaint is the burden of anyone who aims at improvement; but there is a constant undercurrent of something close to despair."

Yet again, for all its contemporary appeal this is not quite right. If one goes back far enough, to Jefferson and Franklin, for example, one finds the very opposite of lament or gloom regarding education. More than the other Founding Fathers, these two men argued passionately for improvements in what was taught at the college level, and they did so with an optimism that I am tempted to call celebratory. At the center of their respective programs lay the idea of Science: science as a means to strengthen the individual through love of truth, discipline, and exactitude; science as a means to advance the nation economically, through technology; science as a requirement for the securing of political freedom through a worldly, informed, and ethical citizenry; science to launch the great experiment of America, whose task would be fulfilling the Enlightenment dream of human perfectibility.

A look at the evolution of curriculum debate since the early days of the Republic shows that science has always been an area of special concern and hope. The contours of rhetoric and image used to promote the study of science have changed dramatically, in fascinating and illuminating ways. Promoting science as an idea or vision has never been very far from selling it as a necessary object of study—or, beginning in the late 19th century, as the guiding intelligence for the very structure and method of education itself. Until quite recently, before the middle part of this century, selling science was bound up with millenarian, utopian ambitions. Indeed, if styles and eras of popularization found their reflection in the classroom, as one might expect, it was not because learning, scientific or not, was always foremost among the minds of educators—certainly not learning for its own sake. It was more because "education" from the very beginning was viewed as a means to make and remake society itself, to form the generations of the future according to a brighter image. And even much later, when ideas of psychology, efficiency, and performance became central, following the advent of mass schooling, the types of "scientific education" that emerged were no less part of this—indeed, they represent nothing so much as its full modernization.

Whether as a subject area or as a teacher, then, science has long been a nucleus for education in America. If the former role dominated during the first century of the Republic, this was perhaps because the latter role had to wait until science acquired professional standing before it could become the source of "facts" and "principles" for general institutional use. In both cases, however, the images of science held by various sectors of society—the general public, politicians, educators at all levels, and scientists themselves—were never separated from each other, nor from the political realities of the moment, and formed the broader context within which the merger of science, education, and the future took place.

Something more, perhaps, should be said about the larger American view of education in particular, for it is unique, not the least when compared to Europe. And the role of science within the American educational world cannot be fully grasped without some mention of the beliefs involved. To begin with a contrast, in 19th-century England, Germany, and France, "to educate" meant primarily to train people to assume specific social roles. According to the European view, education existed to teach people to accept their station in life, or—in a very few cases—to prepare them to assume a higher station. In these countries the curriculum was intensely classical; moreover, it ended with an extremely difficult series of exams (e.g., the *Baccalaureate* in France and the *Abitur* in Germany) meant to restrict entry into higher level positions of sociopolitical power, especially the professions and government service. Napolean had created the model by bringing the French school system completely under government control in the early 19th century. His example was imitated by Prussia, whose leaders traced their wartime defeat to the French system of education. From then on, teachers and professors became state employees, political appointees in a sense. Their work, though "classical" in terms of subjects taught, was drenched in political ideology: At the university level especially the mission of training future leaders who would uphold and further the aims of society as defined by those who yielded power was always clear and palpable. No significant dissent from this mission was ever tolerated. Nor did the sweep of romantic sentiment or the failed revolutions of 1848 have any important impact on the link between the state and education. Indeed, the classical curriculum itself was never contested, and for a particular reason. Graduates were urged to view themselves as standing at the peak of civilization itself, as those who had absorbed and therefore nationalized the most powerful and enduring wisdom of "the ancients." In concept as in practice, "education" was rarelly viewed as a system with the potential to challenge the given order of society—and even when it was, by rebels such as Rousseau, the demands for change were reserved for young children, bearers of nature and fragile innocence. Regarding all others, university students in particular, little was said. Goethe's Werther and Stendhal's Julien, in their quest for individualism, remained true to their Homer no less than the courtiers of a century before. Higher learning in Europe of the 18th and 19th centuries was assumed to be an essential means for creating leaders, a selective system for controlling ambition and guiding social relations. Once in place, this system came in turn to define the idea of "education." Europe lacked a deep-rooted mainstream belief that equated education either with reform or revolution.

Education in America has always been conceived in a radically different way. Here, from the very beginning, education has been viewed as a transforming experience, as something akin to a powerful religious conversion. Bernard Bailyn (1960), among others, has tied this to such things as the missionary interests of the early colonists and the forbidding hardships of wilderness life, for which there were no "prescriptive memories," no body of "reliable lore or reserves of knowledge and experience" (p. 22). Life in America also brought changes in family structure, including the rise of the nuclear home, that made institutional learning suspect or unacceptable as the final site for the imparting of communal civilization. Well before the founding of the Republic, aspirations of religious community, of change, stability, and reform—and, to be sure, of simple survival—had all been invested in the child. More than the progeny of nature, children stood as the carriers of the immediate future, which was always uncertain. They were the means to create this future, to unburden it of the insecurities and failures of the past.

Whether in Puritan New England or in 19th-century New York, formal education came to represent a replacement (and victory) by the state for the family's monopoly on the transmission of culture. It rose and expanded as the total power of the family declined. And thus "education" itself came to assume the child's profound moral and cultural innocence as well as his or her capacity for being trained to live in and shape a new world. Education was placed both at the origin of society and against it, the means to make it and to make it over. "Mothers and school-masters plant the seeds of nearly all the good and evil which exist in the world," wrote Benjamin Rush, with typical mythic reduction (*On the Occupation of the Teacher*, 1790; see Runes, 1947, p. 114). "Its reformation must therefore be begun in nurseries and in schools." In thinking this way, early American reformers understood something critical: that a curriculum for learning, for the "transfer of civilization," bore within it a blueprint vision of society, of social relations, and of the individual.

The primordial myth of American education is therefore a simple one, but also an enormously commanding and continuous one. Indeed, in its equation of learning with the possibility for change (whether conservative, liberal, or radical), with the prospect for individual and social betterment, it is perhaps the

single, most important of all American myths. Because Americans conceive education as a force for change, it has always
been hotly fought over by any number of competing interests.
For groups of all kinds, education has been a means to establish
legitimacy, a site where one's voice could be inserted into the
mainstream. For the individual, too, it has been a place where
one might find the means to climb out of history, beyond the
bounded locale of background. And for leaders, of whatever sort,
education has been viewed as the best means to create a stronger
economy, a more perfect polity, a greater society.

From the beginning, then, the American view was much
too expansive, much too millenarian, much too open to a
confusion of claims, to ever allow for wholly centralized state
control on the French or Prussian model. No state examination
system capable of creating a social elect and, at the same time, a
mass of unredeemable *Volk,* was ever instituted. As Brock (1990,
p. 951) has noted, higher education, in particular, "never came
to confer status upon individuals [in the United States] in the
way it always demarcated elite groups in Europe." Rather, other
more indirect means were needed to do this: for example, status
associated with particular schools, with educational level, with
subjects of study, with specific achievements. But no such
scheme as existed in Europe was ever proposed for the United
States. Even the idea of a great national university—as an
addition to, not a replacement for, an already diverse system of
private colleges in early and mid-19th century—never got very
far before failing altogether (it became, instead, the Smithsonian
Institution).

Because the United States has never had a centrally controlled system of education, it has always lacked a centralized
image of "Americanness," one taught by our system of learning.
Arthur Schlesinger Jr. (1992), among many others, calls for a
return to a consensus type of education, one that might help
reestablish "the vision of America as melted into one people . . .
a unifying American identity" (pp. 14, 17). But such an education never existed in this country. History proves it a fantasy.
History also reveals that the "melting pot" idea was merely a
useful ideological weapon, forged as part of the effort to further

the cause of assimilation during a period of massive immigration (the early decades of the 20th century). Indeed, history reveals this "melting pot" image to be merely one among many others, of which the counteridea of the "mosaic" (discussed by John Dewey and others) figures prominently and with profound suggestive contrast.

Defining who is, or should be, "American" has never been a stable idea within our halls of learning. Those who hold this type of illusion, and there are many today, want to claim something essential, something immanent; they want to lay claim to the past itself, as a time when ideals about learning were more shared, less "tribalized," full of faith in the possibility of one people. It is more than an illusion or fantasy, therefore. It represents an important, traditional plea for the transformational power of learning. It exaggerates; it begs; it demands and threatens. It does all these things because of a committed belief in the impossible—namely, that education alone can overcome all social divisions, even the most profound and bitterly held ones, in a single bound, that it can build a single identity out of a myriad of contesting interests and backgrounds. Schlesinger's plea, then, is a perfect and timely example of the mythologic contours that continue to surround the idea of "education" in this country.

Over the centuries this mythology has never waned. In many ways it remains the only fundamental aspect of education in the United States that has never been seriously challenged. Through everything, the ideal of the willing and malleable citizen, thirsting for education, capable of radical change and productive growth, has remained intact. Within the borders of myth, this individual is itself an institution, certainly of belief. In his or her presumed capacity to absorb and be molded, in the promise of determined freedom, this individual is the foundation on which so much depends and to which so much returns. According to this myth, the student, of any age and at any level, is always a child. Through education he or she is prepared to confront the world and to transform its wilderness. The school, in turn, is always parental, a trusted guardian of present and future. In the beginning, of course, schools took children away from their homes and parents, who then demanded a learning created in their own perfected image. "When the Children are capable of it, I take them *alone*, one by one. . . . *I pray with them* in my Study and make them the Witnesses of the Agonies, with which I address the Throne of Grace on their behalf." So said Cotton Mather, on the subject of *Some Special Points, Relating to the Education of My Children* (1705). He offers an enormously suggestive image, powerful and disturbing; and it is not without its constant echo in later times.

But we should return to science. Where has it fit into this mythology of learning? Some flavor of its changing role, in fact, might be gleaned from a brief comparison. One of the earliest delineations of the value of science to America comes in a letter written by Thomas Jefferson, postmarked March 24, 1789, from Paris.[1] Responding to his friend, Harvard professor Joseph Willard, who had just informed him of an honorary degree bestowed by the college, Jefferson felt moved to speak of the future Republic in this way:

> What a field have we at our doors to signalize ourselves in! . . . It is for such institutions as that over which you preside so worthily, Sir, to do justice to our country, its productions and its genius. It is the work to which the young men, whom you are forming, should lay their hands. We have spent the prime of our lives in procuring them the precious blessing of liberty. Let them spend theirs in shewing that it is the great parent of *science* and of virtue; and that a nation will be great in both, always in proportion as it is free. (in Peterson, 1984, p. 949)

Exactly two centuries later, one finds that Jefferson's concern over issues of knowledge and freedom has been replaced by an establishment rhetoric far different in character and design. As "the scientific" has expanded in terms of real-world power—no doubt far beyond what Jefferson might ever have imagined—it seems to have shrunk in its ability to evoke hope. Or, as pleaded recently by the National Endowment for the Humanities (NEH), science must be merged into a purpose of more anxious intent:

> To the task of learning about oneself and the world, a required course of studies can bring needed order and coherence. . . . A core of learning also encourages community, whether we conceive community small or large. Having some learning in common draws students together. . . . When that common learning engages students with their democratic heritage, it invites informed participation in our ongoing national conversation: What should a free people value? What should they resist? What are the limits to freedom, and how are they to be decided? (1989, pp. 11–12)

Jefferson, loyal son of the Enlightenment, turned his eyes upward: Generous and demanding, forward-looking, filled with optimism, he retained throughout his life the belief that knowledge and liberty were blood relations, sharing bread under the same roof. One notes that Jefferson's vision—like that of the Enlightenment as a whole—did not, in its time, extend to women or to blacks. It thus shared certain limits of its age, even while attempting to break with others. Yet one feels, too, that such limits stood in contrast to the inherent generosity and expansiveness of this vision, which could easily accommodate all groups within American society. For Jefferson, American liberty would give birth to an American form of learning, an American science, and a truly American kind of virtue. Perfection is within man's grasp if he but apply himself to the quest for these things, which extend far beyond the realm of politics, into the most elevated domains.

For the NEH, by contrast, the view from the bridge looks downward and back. Complaint and despair are now relevant. One still hears, as clear as ever, the belief in education as a means for conversion, for rebuilding the social fundament. But the faith is protective, not expansionist. The "field at our door" has shrunk tremendously. Instead of reaching outward with festive ambition, we must pull back and defend what is left, a sacred nucleus. The NEH makes its call for obedience to "heritage," for allegiance to "community" as an idea and image for binding all things to a controlled center. The learner is no longer at the nexus of things; he or she now stands outside, waiting permission to enter: As student, he or she must be "invited" in.

Such differences are momentous. Indeed, they seem irreconcilable. Jefferson's idea of the student is a magnificent investment. He or she must be someone eager to throw off "the desponding view that the condition of man cannot be amelio-rated, that what has been must ever be, and that . . . we must tread with awful reverence in the footsteps of our fathers" (*Report of the Commissioners for the University of Virginia*, 1818; in Peterson, 1984, p. 462). Such an idea, given contem-porary expression, could only appear threatening (certainly anti-communal) to those who today share the NEH position, for whom the student is much less a creator of culture than its mere vessel or querulous servant.

What, then, of science, in particular? To Jefferson, clearly, the learning of things scientific would pave great and broad avenues to empowerment, of many required kinds.[2] Science should be studied for the sake of creating a citizenry capable of improving itself beyond the known limits of independence and responsibility, for negating parochial despotism in any form, for inspiring "honesty" and "social worth" at all levels of experience and action. For the NEH, on the other hand, the learning of science has a different tone of purpose:

> Our ability to make everyday decisions wisely is diminished when we do not comprehend scientific principles and the technologies built upon them. . . . To the task of understanding [vital questions] and knowing what we can about ourselves and the human situation, the sciences bring creativity and imagination as well as rigor and preci-sion. (1989, p. 43)

What a limp and paltry rhetoric is this, when placed beside the words of Jefferson! Not merely uninspired and flat, it seems unstable as well, an unconvincing recitation of platitudes and vague invocations. Nothing of the vigor of Jefferson's *interest* can be found here, nothing of his faith *in* and hope *for* the student.

As I intend to show, this is no accident. It has everything to do with what has happened to education in the meantime, what has taken place in terms of theory, methods, motives, and much more—a great deal of which has been founded on ideas of "science" and "the scientific." Loss of the relevant idealism is the result of many histories at once. Those that surround curriculum debate and the larger images of "science" are paramount among these, but they are not the final answer. Rather, these areas have most often acted as focusing media. Science has acquired the type of rote justification provided by the NEH due to the efforts of scientists themselves and to those of conservative educators in this century, whose program for unifying and standardizing learning has emerged whenever social change has prompted greater openness in higher education. Part of this program has been to promote certain ideas of "civilization" or "culture"— presumably embodied in the Western humanities—as the necessary core for all higher intellectual experience. Correlative to this has been a two-pronged consideration of science: first, as a general area of endeavor separated from the humanities, concerned with "facts" and "principles" about the material world; second, as a collection of individual disciplines that can be taught in a fashion similar to that of the humanities—that is, by means of a "great discoveries" and "great men" approach that effectively brings science well within the larger idea of "civilization" reduced to a series of peak minds and moments (Western, to be sure) that stand beyond criticism. This bimodal view, which both holds science apart yet absorbs it, is one that scientists themselves have often held, with some amendments to be sure. One might well note, even at the outset, that it is a view intended, at some level, to keep our expectations of what the sciences can do and be, as subjects of critical study, within strict status-quo limits.

But for now, other questions should be given precedence. For example: What history links and separates the beliefs offered by Jefferson and the NEH? How has the rationale for the study of science changed over the intervening two centuries, and to what were these changes related? What role, finally, has science itself played, as an area of expertise, in the making of American education?

CHAPTER ONE

Science and Democracy:
Emerging Trends of Faith

Nearly every writer and orator of the early Republic called for the establishment of a distinct and distinctive American form of education. Major statesmen eagerly offered plans for the founding of a national system of schools or colleges. As concept, as philosophy, and as metaphor, science (and technology) was central to nearly all such plans. Jefferson and Franklin put forth programs for actual scientific study, while others such as Benjamin Rush hoped for an education that would turn men into "republican machines" (Runes, 1947, pp. 87–96), and Thomas Paine spoke of the new country as embodying the principles of a Newtonian universe. Everywhere one looked, one found "science," both guarantor and emblem of the "national experiment." As Lawrence Cremin (1980) has said,

> They [the early reformers] urged a genuinely useful education, pointedly addressed to the improvement of the human condition. At its heart would be the new sciences, through which citizens might come to know the immutable laws governing nature and humankind and on the basis of which they might build a society founded on reason and conformity to moral truth. . . . Through the systematic application of science to every realm of living, they would learn in countless ways to enhance the dignity and quality of their daily existence. (pp. 3–4)

Something else lay behind this grand emphasis: the desire to found a singularly *American* kind of education, antithetical to that which had emerged from and long supported the reign of tyranny and feudalism in Europe. Science was assumed to be a realm of thought and endeavor both practical in nature, therefore supportive of personal and public independence, and republican in spirit, therefore democratic. It was a realm therefore that challenged and reformed Old World class divisions and corruptions.[1]

FRANKLIN, JEFFERSON, AND THE CURRICULUM

Up to that time two types of secondary schools existed, neither of them public. The first were the so-called academies. These were small, local, private institutions, either "Latin" in focus, with the sole function of preparing young men of wealthy families for college, or offering a more varied curriculum intended to train graduates for business or teaching. The second type, were the "charity" or "pauper" schools, supported by local governments for the sake of orphan or indigent children. Charity schools served a variety of social purposes: to uplift the poor morally, to instill discipline and respect for authority, to reduce the possibility for crime, and to provide some measure of comfort. Such schools were the closest thing to public education that had yet been conceived—but they were a long way from deserving that title in any but a metaphorical sense. Imagine, therefore, the radical nature of a plan such as Jefferson's, unveiled as early as the time of the American Revolution, involving an entire system of state-supported popular education, beginning with district elementary schools and progressing all the way up to a grand State University of Virginia! Nothing so comprehensive and centralized had ever been advocated in America before, not even by Franklin. Jefferson's plan was intimidating in scale alone. And what type of subjects did he propose for study? What type of "Republican mind" did he, and Franklin, have in view for the future, particularly at the higher levels?

The curriculum devised by Jefferson and Franklin departed sharply and deliberately from the standard European model of higher learning, typified for example by Oxford or Cambridge. Their deviation was a simple one: In addition to the usual subjects, Latin and Greek above all, with a smattering of modern languages, history, and so on, they added courses in science, both "pure and useful." In his proposal for founding a model institute of higher learning in Philadelphia (the Constitution of the Public Academy, 1749), Benjamin Franklin called for the following offerings:

> For attaining . . . great and important advantages, so far as the present State of our infant Country will admit, and laying a Foundation for Posterity . . . an academy for teaching the Latin and Greek Languages, the English Tongue . . . the most useful living foreign Languages, French, German, and Spanish . . . History, Geography, Chronology, Logick and Rhetorick, Writing, Arithmetick, Algebra, the several Branches of the Mathematicks, Natural and Mechanick Philosophy, Drawing in Perspective, and every other useful Part of Learning and Knowledge. (in Labaree, 1959, p. 423)

This basic plan was repeated by Franklin on many other occasions, for example, in his well-known *Proposals Relating to the Education of Youth in Pennsylvania* (1749; see Labaree, 1959, pp. 397–421) and in his dedicatory speech, *Constitution of Phillips Academy* (1778). In every case he argued that "the first and principal object . . . is the promotion of true piety and virtue," by which he meant that education should be a source of personal and therefore social improvement. He placed emphasis above all on those subjects that would be "most useful," on practical study that would be both "pleasant and agreeable" instead of "painful and tedious," and on a curriculum that would yield such students as might be "fitted for learning any business, calling, or profession."

This does not mean that Franklin was a strict utilitarian. On the contrary, what he wanted was an enormous, unprecedented opening up of the curriculum. As Bailyn (1960) puts it:

> Indeed, he did expect education to be useful, as who did not; but his revolution consisted in the kind of utility he had in mind. He wanted subjects and instruction that trained not for limited goals, not for close-bound, predetermined careers, but for the broadest possible range of enterprise. He had no argument with the classics as such. What he objected to was their monopoly of the higher branches of education which denied the breadth of preparation needed for the open world he saw. (p. 35)

In other words, it was not merely a new student that Franklin hoped for. What he foresaw was a new type of person, fitted for life in diverse ways, provided with a range of possibilities that had not previously existed. In "education" he saw directly and immediately a vision of the Republic itself, freed from feudal apprenticeship to Europe. Education, in short, was linked to ideas of revolution. To Franklin, Jefferson, and other early proponents of a new type of learning, a deeper aim was providing material for a new brand of citizen. To them, as Paul Goodman once said, " 'Citizen' meant society-*maker*, not [merely] one 'participating in' or 'adjusted to' society." The making of society was "their breath of life" (1964, p. 24). Education in the sciences, then, was a critical element in this effort to make and to build.

It is interesting, in this respect, to compare Franklin's proposals with Jefferson's curriculum for the University of Virginia, offered 70 years later in 1818. Here, ten study areas are laid out, of which fully half are in the sciences. The hierarchy, too, is revealing: After "languages, ancient" and "languages, modern," the next five proposed study areas are all devoted to science ("mathematics, pure"; "physico-mathematics"; "physics, chemistry, mineralogy"; "botany, zoology"; "anatomy, medicine"), with the last three areas being taken up by "government, political economy"; "law, municipal"; and finally a hodgepodge category of "ideology, general grammar, ethics, rhetoric, belles lettres, and the fine arts."

Jefferson's program differed from Franklin's in that it was less focused on "the useful arts" (though it included many, from agriculture to military fortification), yet gave central stage to the sciences as a whole. For Jefferson, it would seem, the study of science should precede or guide the study of politics, history, law, indeed civic knowledge as a whole. This was because, in contrast to Franklin, for whom science was important in proportion to its degree of "usefulness," Jefferson more directly shared the Enlightenment ideal of science as representing a kind of grand and utopian teleology, set in motion by humankind's fundamental desire for progress. This meant science as a combination of intellectual, moral, and spiritual methodologies whose ultimate end was nothing less than the perfecting of individuals and society as a whole. It was within this frame, then, that freedom for Jefferson could qualify as "the first-born daughter of science."

SCIENCE AND ADVANTAGE

During the period spanned by these two curricular proposals—a period that both preceded and included the Revolution, its aftermath, and another war with England—science came to be attached to an enormous range of republican virtues and values. The language surrounding these ideas and feelings has, one might argue, set the basic scheme of debate down to the present day (though of course in varied form). For this reason, it is important to list the principal advantages science was thought to offer. Five, in particular, are notable:

1. *Pragmatic Advantage.* Science, in the form of Franklin's "useful arts," was promoted as a subject of learning for the sake of work, for improving individual and collective productivity and wealth, for "fitting" one to an "advantageous life." Farming, the prime source of America's wealth (it was, along with the supply of raw materials such as lumber and coal, the pillar of the American economy) and the symbol of the ideal, independent American citizen, was a particular target of this rationale. Jefferson, among a host of others, often argued that "just weight" be given "to the moral and physical preference of the agricultural, over the manufacturing man" (letter to Jean Baptiste Say, 1804; see Peterson, 1984, p. 1144). Soon, however, this "manufacturing man," whatever his presumed limitations, saw the importance of a technical education as well. A growing number of scientists, mill owners, and inventors began to lobby strongly for science education on the basis that it would enhance industrial commerce and therefore opportunities for employment and personal economic advancement. John Adams quickly added his voice to this lobby (he would argue on behalf of scientific education all his life), and in so doing, revealed an investing of science with the fundamental virtues of the work ethic that

would become so central to American mythology in the next century: "Manufactures," he said, "cannot live, much less thrive, without honor, fidelity, punctuality, and private faith, a sacred respect for property, and the moral obligations of promises and contracts" (in Gutman, 1977, p. 5). A government "decisive, as well as . . . intelligent and honest" could promote these virtues, but only if it took science to heart, making it the focus of its vision, and of a "new education" for the American citizenry (p. 5).

2. *National Advantage.* The training of engineers, architects, inventors, and industrialists was promoted by many leaders for purposes of national economic independence, for freeing the new nation from a widely recognized and humiliating dependence on foreign expertise and foreign goods. This new "native corps" would promote fuller employment at home, encourage a higher degree of loyalty and dedication (workers would more gladly devote themselves to productive labor under American experts, rather than foreign ones), and would therefore further both trade and national identity. Encouraging education in science/technology would directly aid domestic manufacturers, and thereby provide a vital and long-term defense, a kind of solidification, of the republican venture itself.[2] Indeed, in the post-revolutionary period, education as a whole came to be viewed as a nationalist enterprise, such that proclamations against families sending their children to Europe for study became frequent and were even enacted into law in several instances (e.g., Georgia mandated in 1785 that any persons under 16 who had studied "under a foreign power" for three years or more would, upon their return, be treated as "aliens" for an equal amount of time [see Knight & Hall, 1951, p. 93]). George Washington, near the close of his second term as president wrote, "It is with indescribable regret that I have seen the youth of the United States migrating to foreign countries, in order to acquire the higher branches of erudition, and to obtain a knowledge of the Sciences" (Fitzpatrick, 1940, p. 149). Noah Webster in a May 1788 issue of the popular *American Magazine* went still further: "Nature has been profuse to the Americans, in genius. . . . If this country, therefore, should long be indebted to Europe for opportunities of acquiring any branch of science in

perfection, it must be by means of a criminal neglect of its inhabitants" (Knight & Hall, 1951, p. 95). Educational independence—with science in the forefront—was thus seen by many leaders as commensurate with national independence, with liberty itself.

3. *Moral Advantage.* Science and "scientific thinking" were viewed as encouraging the betterment of individual character. Honesty, patience, discipline, observation, civility, and industriousness—"virtue," in other words, in all its Protestant forms—were frequently mentioned as the probable, even the inevitable, result of a solid grounding in the sciences. Interesting to note here, as discussed by Kasson (1976), is how in the American mind Britain and Europe became associated with "luxury and dissolution," "extravagance," and the like, for all of which the "firmness" and "progressive spirit" of science could act as an antidote by gaining for the individual a more solid control over "the power of instinct."[3]

4. *Republican Idealism.* In this category is to be counted the association, on the level of spirit and sensibility, between science and such things as liberty, inevitable progress, and human perfectibility. This, as I have said, came largely out of Enlightenment belief, but it was adapted to the republican cause in particular. Science meant revolution, new forms of openness, exploration, a glorious victory over the prejudices and tyrannies of the past. The new nation, America, was to be the site of a great intellectual quest, where liberty would nourish learning and the advancement of learning would promote greater liberty.

5. *Pastoralism.* This is perhaps the most complex and interesting category of all. Even before the time of the Revolution colonial American thinkers and writers had created the vital myth of America as "Nature's nation."[4] Within the old capitols of Europe lay the rotted fruits of inequality, squalor, indolence, confinement; the New World, by contrast, was characterized by an Eden-like fertility, on an unspoiled and magnificent scale. Certain forms of goodness and vigor, both personal and social, were believed to derive from direct contact with such a unique natural environment. A special love of country, dedication "to the land," a willingness to think large, to work hard and generously for self, for one's neighbor, and for communal

welfare too—these were the pieties often cited. During the early and mid-19th century, this myth evolved into the concept of an "American nature," scientifically distinct from that of Europe. Thomas Say, a well-known entomologist of the early Republic, rejected the older Linnaean system of taxonomy out of hand. As Stroud (1992, pp. 3–4) points out, his "intensely nationalistic goal of establishing the authority of American scientists to name and describe the flora and fauna of their own country placed him squarely in the mainstream of postrevolutionary thinking."

More broadly, certain scientific subjects—in particular, botany, zoology, geography, and geology—were recommended for both academic and popular (self-) learning in order that citizens become directly acquainted with the "native wonders" and "natural lessons" of the Republic. Well into the 19th century it was common to hear educators, secular and Christian alike, argue for study in these areas on the basis that American values and virtues lay written in the native landscape, that size, grandeur, hardiness, and the "lure to greatness" were all to be culled from a familiarity with "the face of God's chosen land." The great landscapes painted by America's best artists in the years between 1830 and 1880 testify to the power of this pastoral myth (see Novak, 1980).

Thus the demand for "clear powers of observation," revealing of the national superlative, was equally shared between art and science, both of which partook of the literacy of "Nature's nation."[5] In this way, then, the pastoral ideal in revolutionary America later became transformed into an argument for scientific study as a kind of political nature worship.

These five categories, of course, stand as artificial divides in what was a heady and unmitigated flow of overlapping favoritisms. Moral, political, and practical sentiments aimed at the heroification of the Republic through science (and therefore, science through the Republic) tended to join like separate tributaries of a single vast current. Those who advocated scientific study commonly spoke of the need to promote such things as industry, thrift, temperance, and self-denial, to battle against any erosions, perceived or foreseeable, in character,

learning, and nationhood. Most leaders at the time, being intellectuals and writers as well as politicians, scarcely made any
real distinctions between these latter areas. But the promotion of
science also came because of something inherent in the very idea
of "education": namely, that this process was formative in the
most profound ways, not only for cultivating "the germs of
intelligence, uprightness, benevolence . . . that belong to all"
(as Horace Mann later put it; see Cremin, 1957, p. 12) but for
enabling the individual American to go even beyond the historic
limitations of his nature, to achieve something of the genius that
was to be America. Science, as cupbearer of reason, could not
but be central to such a way of thought.

All in all, then, "science" was a grand and yet vague all-
encompassing idea. It was like a flag picked up and waved by
many eager hands. Jeffersonian idealists could use it to argue for
a learned citizenry. Conservatives, like John Adams, promoted
the value of science for encouraging a mixture of self-restraint
and inventiveness. Radicals like Paine, schooled in Scottish
"common sense" philosophy, could find in it the proof of a
deistic universe that denied any final authority for the church
and confirmed the evil of aristocracy in any form. Manufacturers, meanwhile, themselves interested in profit, could expound
on science as the key to progress and economic independence.
Still others might speak of its use for the building and rebuilding
of the material nation itself (with which such disciplines as
physics, chemistry, engineering, and medicine were correlated).
The images surrounding "science" therefore were very often the
same as those enveloping the idea of the Republic, both physically and spiritually.

PHILOSOPHICAL AND CURRICULAR TRADITIONS IN AMERICAN EDUCATION

Within the bounds of these multiform rationales, guiding them
forward in the area of actual educational policy, were three
traditions, one ancient, the other two relatively new. These
traditions have remained in place ever since; indeed, their
visions of learning, of the power and place of education, and
therefore of the ideal individual and his or her collective society,

are more evident today than they were back then, for a host of reasons. Two of these traditions, what I call Academism and Practicalism, emerged by the early 19th century and assumed much the same form they have today, remaining part of the educational landscape ever since. The third tradition, Reformism, is different in crucial ways. Its values have been revolutionary, even radical (when viewed from a status-quo perspective), rising out of active disillusionment during times when a general shift occurs away from obedient faith in given institutions toward a more person-centered, utopian view of society. As such, it has punctuated rather than led the history of American education. Yet its ideals are in large part the most democratic of all.

What follows, then, is a brief description of each of these central traditions.

Academism

Academism can be defined as the oldest educational philosophy in modern Western culture. It can be correlated with the basic mode of "ivory tower" or Old World "liberal arts" learning. This tradition reached maturity in England and France during the era of growing empire (the 16th and 17th centuries).[6] It grew out of a system of higher learning originally intended to strengthen and perpetuate monarchical government and its elite servants, by training—psychologically, linguistically—a stratum of "higher" individuals. In the beginning, it was derived directly from medieval educational structures, in which learning was based on the *auctores* system, the reading and close study of canonical authors celebrated as the providers of timeless, fundamental wisdom. The "great ideas" educational philosophy was based on intimate contact between the individual mind and the "great book"—the tutor–pupil relationship, in a transferred setting. Granted, this was an ideal, not so often achieved (much of medieval learning, based as it was upon strict memorization, more closely resembled manual labor). But in the courtly societies of 16th-, 17th-, and 18th-century Europe, when the universities were filled with the sons of wealthy men, and when the idea of statesmen intellects, whether destined for politics or scholarship, came to the fore, the feeling of a need to condense and have at hand the heroic wisdom of heroic times (classical Greece, imperial Rome) became fully attached to the "liberal arts." That a committed and long-term opposition to this way of thought existed, including the likes of Frances Bacon, John Locke, David Hume, and others favorable to science, did not much matter. Indeed, the well-known quarrel between "ancients and moderns"—between those favoring belief in the supremacy of classical knowledge and those who saw modern knowledge, caked by science, as far surpassing that of Greece and Rome— had little or no effect whatsoever on the basic philosophy of European university teaching. It took place outside this setting and did not at all interrupt the continued production of gentlemen-scholars, whose training bound them irretrievably to existing systems of patronage.

In social terms, canonical works of literature and philosophy were the vessels by which Greek learning, in particular, had been appropriated to the "European" as the embodiment of Western (read: advanced) culture as a whole. That a great deal of this learning had come to Europe via translation from Arabic sources in the 12th and 13th centuries was largely, if not totally, forgotten. In fact, the forgetting of debts to Islamic civilization was commensurate with the adoption of the new knowledge into the old *auctores* system and its use in the building of the new universities after 1200. These were originally set up throughout Europe in imitation of a guild-type structure, then developing within the major trades. As Lindberg (1992) notes, they aimed at "self-government and monopoly": To accomplish these ends meant to ensure a restricted collegiate community. This involved the setting up of a structured series of exams and a tutorship system, and depended on a high level of patronage from kings, popes, and princes, all of whom came to view the universities as "vital assets," centers of both spiritual and worldly power, which needed to be "carefully nurtured." Elements that challenged or denied the preeminence of such patronage, such as a Muslim *auctore*, were gradually eliminated. The idea of "higher" with regard to this kind of education, as well as its selective, earned character, thus came to be institutionalized on several levels at once—behind every claim for an "elevated pursuit of learning" lay a very concrete, real-world structure of political force and protectionism, one with its own motives for self-elevation.

By the time of the Renaissance, the university had become the major institution of learning, with its guild-type structure firmly in place (as it remains, in large part, today). Socioculturally, it did not create a division between knowledge and life, but rather a contouring of knowledge for specific types of elitist life and practice. These, one could say, were concerned with leadership in the imperial, colonialist period of national expansion. Education was aimed at training statesmen in arts and politics, those for whom the image of select "geniuses"—monarchs of the mind, so to speak—was of obvious utility in providing both a model and a promise of induction into hallowed superiorities. Over time, the indigenous political-cultural contents of such an education came to be denied on the basis that "higher," "universal" truth was being served (still later, this became a basis for the idea of "academic freedom"). But the real essence has always been to produce a kind of intellect that embodies ideas of titled superiority and entitled stewardship, an intellect not so much capable of critiquing as personifying forms and assumptions of "authority." For this reason, the Academist tradition has of necessity granted privilege to the curriculum first (canonical wisdom), the teacher second (as the representative of such wisdom), and the student last (as novitiate). Often enough, in its various arguments for reform, it has left out the middle term altogether: Such is the presumed power of a canon that it requires only a messenger, not a true teacher.

Not ironically, science was long given a reduced importance, a more-or-less secondary role, within the limits of this tradition. The reason for this is fairly straightforward. The ideal individual, encoded within the "liberal arts" curriculum, was someone who demonstrated maximum powers vis-à-vis language, its forms and uses, both written and verbal; such, after all, was the substance of leadership at the time in art and politics.[7] With the exception of natural philosophy, science did not fit well into this paradigm because it was linked with the "practical arts." Indeed, for most of its history, science grew outside the university system altogether. Only later, once scientists had themselves taken on the robe of higher truth—through such notions as "objectivity" and "disinterest," therefore claiming complete removal from the rest of culture—was acceptance forthcoming (and then only slowly).

This, then, defines the second and more modern tradition of Academist belief, this historical accommodation. Such notions as "dispassion," after all, were by no means unique to science. By claiming them as their own, scientists were adapting an existing rhetoric to a new purpose. Their success was enormous, their knowledge seemed to prove such claim making. Very soon any connections to the older Academism were lost or ignored.

All the colleges founded during the prerevolutionary period in America were based directly on this older tradition, imported from England. To be counted among these are Harvard, Dartmouth, King's College (later Columbia), William and Mary, Yale, and the College of New Jersey (later Princeton), among others. These institutions, indoctrinated into the sensibility of their origins, plainly resisted any broadening of scientific curricula after the Revolution and into the 1820s. At that point, certain "pure" science courses were established or broadened; it required another half-century or more, and the looming success of the new research university, before anything approaching a conversion to the new Academism was grudgingly established.

Practicalism

Practicalism, which has both opposed and buttressed Academism, was touched on above in my discussion of Franklin's ideas about education. Simply put, it valorizes the notion of learning as a source of practical advancement, whether for the nation, for industry, or the individual. Often, this has meant promoting the idea that education should be directly connected with the life and needs of ordinary society—by which is usually meant economic opportunity for working-class and middle-class people. Those who believe in and promote the practical view have always put forward the notion that school should be a place of vocational training, where one is prepared for practical life. Indeed, they argue that the prime purpose of a public school system is to train students for work and careers.

Alternately, however, the idea of "practical benefit" has stimulated demands of a more general and (often) unfocused nature, for example, that higher learning should serve "the nation" or the "national interest" with regard to such things as economic competitiveness, military power, or international prestige, and that it should contribute to, and even be guided by, the needs of U.S. business and industry. As a result, Practicalism with regard to scientific education has usually been directly oriented toward technology and the so-called "fruits of science." Yet in its nationalist mode, it has frequently made use of Academist images of science, particularly during eras when these images have commanded general belief, arguing that "pure science" will always be in some way American and will serve American interests, when conducted within American borders. At times, this tradition holds within it a view of the nonelitist individual, the "average person" (however defined) whose station in life can be advanced by certain types of learning. Yet when it comes to nationalist ambitions, to goals like competitiveness or technological supremacy, the demand has tended to focus readily on expanding the number of "experts" in society, and therefore on the need for an intellectual plutocracy. With regard to scientific education for industry and "the national interest," starting in the latter half of the 19th century Practicalism frequently conceived the individual in terms of "manpower" or "human resources," that is, as a type of commodity to be employed for larger purposes—commonly those of profit, imperialism, and international power.

With regard to education, therefore, Practicalism has promoted popularization, opening the doors to larger numbers of students. It has not been aimed at reforming or bettering society in any large sense, but instead has wanted to reform education in a manner that gives added strength to existing forms of institutional authority, and that develops and empowers new ones. For these and other reasons, this tradition has also tended to elevate both the curriculum and the teacher above the student. Its model graduate is, to put it simply, a soldier of knowledge—someone who will "apply" his or her training or expertise toward a combination of personal advantage and institutional purpose.

Reformism

Reformism, the last tradition, has tended to view education very differently from either Practicalism or Academism (although it has often included the latter's images of science). The truest proponent of Reformism, and in many ways its historical spokesperson, is John Dewey, who viewed "education" as a broad range of activities in society and who saw school as a living experience, one whose fundamental purpose was to connect life within the classroom to life outside in the everyday world. (In fact, as Dewey often noted, there is always a connection in some sense; education always reflects the society in which it occurs: Academism reflects the continuing class structure of modern society, Practicalism reflects both the struggle between classes for economic equality and the efforts of industry and government to control that struggle.) To put it another way, Reformism has wanted to take this connection in hand, give it form within the school, and direct it toward social change for a more democratic way of life.

Philosophically, this tradition is the exact inverse of Academism: At the top is the student, whose needs with regard to learning-as-a-process always come first; then comes the teacher, who educates through interaction with the student (an equal participant); last is the curriculum itself, whose basic contents remain essential but whose form is to be extremely flexible, always open to "experiment" and change. The ideal individual envisioned within this scheme is one who will embark on a path (in Dewey's words) of "education for life," meaning education as a neverending process, without any necessary aim toward specific practical gain.

Reformism is always a form of rebellion against Academist claims in America. It is, perhaps, the inevitable outcome of a clash between American democratic beliefs, which have a leveling effect on institutions, and the world of higher education, in which forms of hierarchy and the sense of superior attainments (whether of refinement and culture or of expertise) have been endemic since the very beginning. Reformism reflects most pointedly a belief that learning represents a source and a beginning, an opportunity for something new and better to be created. At the heart of Reformism, then, is not the ideal individual as "statesman" or "elegant mind," fit to enter the ranks of the chosen, but as empowered "common man," capable of choosing and creating choice with regard to "the great end and real business of living" (Franklin, 1778, p. 78). It is this individual, whose precise contours have proved so difficult to define, and therefore been so amenable to appropriation by any number of interests—from hopeful utopians to conservative government ideologies—that has helped lend this tradition its aspect of experimentalism, its sometime lack of coherence, and its continued search for discovering and (re)shaping approaches to learning that are exclusively "American."

Science has often held a place of honor within the Reformist tradition, from Dewey onward. This place has been a troubled and often vague one, being centered on ideas of "science for democracy" and "science for a better world." But such ideas, naive as they might have been, hark back to Jefferson directly, to his belief in the possibility for a battle against tyranny waged through learning.

In fact, all three of these traditions, together, have a complex and interwoven history of relationship with ideas about science and its educative worth. At some points in the past, such ideas, current among the public and among powerful individuals, have given direct nourishment to one of these traditions over the others (e.g., the "science for national defense" movement of the 1950s and 1960s, which strongly favored Academist concepts of "pure science"). At other times, proponents of one tradition have sacralized certain images more endemic to an opposing tradition (e.g., Progressive reformers of the 1910s, 1920s, and 1930s, who faithfully took up positivist notions of a "detached" and "objective," that is, an Academist, science). Doctrine and dogma have never been lacking in any of these traditions. Nor have fear and hope as their motivating passions.

During the whole of the 19th century, Academism held sway in primary and secondary education, a traditional Academism that favored the humanities over the sciences. Only during the century's final two decades, by which time scientific knowledge had been fully invested with ideas of progress and national power, was science finally given a secure home in the curriculum. The progressive era, meanwhile, saw the curriculum change radically, according to mainly Practicalist (but also Reformist) rationales: Vocational courses, training in personal hygiene, ethical character, and the like, became standard and remained so until the late 1950s, when Russia's achievement with Sputnik encouraged a swing back toward more Academist programs—now, however, with the sciences in the forefront.

For higher education the story is somewhat different. During nearly the whole of the 19th century, while the prestigious elite colleges stood alone at the top of the educational pyramid, higher education in the United States was firmly controlled by traditional Academism. Later, with the rise of the research university, a reformed type of Academist sensibility became more important; this reformed Academism saw in "science" the highest, most removed, uncompromised, and purest form of cerebral work—a work based on contribution, on the forward movement of knowledge, on a belief in "progress" transferred to the intellectual realm. All this was mixed, too, with a growth in Practicalist-oriented training for professional work, found predominantly in technical schools and in the engineering departments of the expanding universities. In our own century, meanwhile, changes of emphasis have taken place; but it does seem clear, indeed all too clear, that in terms of actual teaching, it has been the Academist tradition of an elevated and objective science that has held sway. Only with the coming of the late 1960s and early 1970s did a truly new, Reformist image of science rise to the surface, yielding new voices of critique, new possibilities for content.

Each era was formative in terms both of what it did for (or to) education and also in what it said. The rationale for programs of change or stability, for certain types of curriculum and not others, and for the uses of education, are as central to any study of American learning as the schools themselves. For the 19th century, it is interesting to note how, during almost the whole of this time, there existed a subtle division between two fundamental points of view. One of these emerged to promote the idea of America and democracy through "arts, letters, and politics" (Adams, Hamilton, and other conservatives); the other argued for freedom through "politics, science, and the arts" (Jefferson, Rush, and others). This was not an early version of the so-called two cultures debate, pitting the humanities against the sciences. It was more a matter of emphasis: For those who believed in the benefits of "the classics," science was distinctly less important in terms of the training of sensibility. For those, on the other hand, who valorized progress and associated it with the securing of democracy, the classics were necessary only to the extent that they enabled one to know and understand the past, to learn its lessons, and to enjoy its beauties, but for the future there was science, and thus science for the future. The hostility between these points of view grew up within the academy, not outside it. It was a rivalry born of power struggles, as an older order, medieval in spirit, began to fall before the "modern." But the fall has never been completed, and the struggle has never been resolved. Inasmuch as it continues today, it duplicates in part the debate between Jefferson and Adams, between the possibility for an uplifted citizenry and the need for a mandarin leadership.

Science in the New Republic: Faculty, Family, and the Failure of Idealism

The double view of republican science, involving a composed look backward and an excited look forward, did not last. Despite the multiform claims made for science in the immediate postrevolutionary era, especially concerning its future role in education building, no widespread reforms were initiated for as much as a half-century afterward. The sentiments of Jefferson and Franklin had been shared by many, especially during the early years of the Republic when a national identity, a national consciousness, was being formed. At this time newspapers and other periodicals often carried articles about scientific matters that drew many correlations between patriotic virtue and the knowledge and outlook of science (see, e.g., Beaver, 1971). But in reality both individual state governments and the federal government, and certainly the colleges, continued to remain entirely in the hands of men whose sensibilities had been trained in 18th-century Academist traditions, whether in England or at home— traditions in which science certainly had a role to play, but one whose significance remained minor.

As the mass consciousness of revolution began to fade, as the new country came to face diplomatic, economic, and military crises, and as the idea of "American empire" rose to the fore (especially after the Louisiana Purchase), the pressures for educational reform on a Jeffersonian or Franklinian model faded. As a society, moreover, America was anything but unified or co-

herent. It more resembled a collective of dispersed, semiautono-
mous communities, often unrelated in any consistent way to the
few urban centers that did exist. Writing at the turn of the next
century, Henry Adams (1885) would say of the year 1800 that its
intellectual life seemed "picturesque, but not encouraging" (p.
95). Harvard, and the colleges in general, "resembled a priest-
hood which had lost the secret of its mysteries, and patiently
stood holding the flickering torch before cold altars" (p. 55).

As such, the new nation was vastly different from Europe,
with its centripetal metropolitan cores, its hierarchical city/
town/village structures, and, of course, its great universities. The
needs of most Americans were local and of the most practical
sort. In the generation after the Revolution, these needs reasserted themselves on a daily basis and education was therefore
viewed accordingly by much of the public—not in the grand
terms of individual opportunity to change one's place in society,
but rather as a potential help to one's family and the immediate
community. Often, education became a tool in sectarian struggle. Competing schools were set up by various denominations as
each jockeyed to expand its influence. Any attempts to central-
ize education at the state level were quickly rejected or else
appropriated by such groups. Despite decades of effort and the
prestige of the presidency, for example, Jefferson failed to put
through his own scheme for a public school system in Virginia.
Franklin's academy in Philadelphia became, in the end, a vic-
tim of lack of interest and of angry strife among religious
factions.

SCIENCE AND THE COLLEGES
IN THE NEW REPUBLIC

In such a climate the colleges were largely left to themselves.
Their sense of mission remained firm, but society had changed
around them. The nourishing environment of colonialism, with
its monarchical substructure, was now gone. Prerevolutionary in
their philosophy and curricula, linked to a social and familial
structure harking back to the Commonwealth era, these institutions resisted change, and became anachronistic. Enroll-

ments remained low, money tight, and connections to the life of the new society meager at best.

This is not to say that change was altogether absent in the last decades of the 18th century and the first few years thereafter. Nor was the curriculum itself wholly fixed and inflexible. Even before the period of major reform beginning in the 1820s (when the colleges were attacked for being "aristocratic"), a number of scientific disciplines such as natural history, botany, mineralogy, and chemistry had been added to the curriculum in most schools of higher learning. But in practice these fields were of minor and unstable status. In terms of teaching, for example, science in the post-revolutionary colleges was handled in one of several ways that directly betrayed its minority importance. Often teaching science was a secondary assignment given to professors of mathematics or philosophy. In other cases, it was a seasonal job offered to local doctors who filled in as part-time lecturers. In both instances, the lack of financial support—whether from the federal government (nonexistent at this time), from higher enrollments (equally unlikely), or from wealthy donors (not yet forthcoming)—meant that all fields of a "science" tended to be taught by the same individual, whose position was far from secure on a year-to-year basis.[1]

A few important professorships did exist: In the late 1790s, at the urging of Benjamin Rush, a position of solid standing had been created at Princeton (then called the College of New Jersey) for John Maclean, America's first professor of chemistry. Maclean published regularly, gained stature as an experimenter, and attracted students to the college—thus, in a sense, he provided an early model for the professional academic scientist. Yet his importance was largely based on the fact that he was not an American. He was Scottish, had studied at Edinburgh and Paris (where he met Lavoisier), had taught surgery at the University of Glasgow, and was therefore a well-established academic European before he emigrated to Philadelphia in 1795. He thus brought the cosmopolitanism and advanced status of European science to American soil, and as such represented a rare "catch" (one, nonetheless, whose securing still required the intervention of a prominent leader like Rush). Yet even all this did not permit him to escape the fate of his contemporaries:

Originally appointed to teach mathematics and natural philosophy, he had to bargain his way into gaining permission to offer chemistry and natural history as well. Thus he became, in name and effect, the college's single professor of science.

Nearly everywhere else the situation was far worse. As Guralnick (1975a), a historian of the era, has described it:

> The science professor, to apply the term loosely, was at best a peripheral entity in the collegiate organization. . . . His salary was uniformly lower than that of other professors, and his security such that he easily fitted the classic mold of the "last hired and the first fired." . . . Those who had positions as professors spent the opening years of the century listening to students recite passages from textbooks whose copyright or composition dated from colonial times. Lecturing, or the presentation of explanations not contained in the textbooks, was a comparatively new and little used technique. Generally unassisted by laboratory apparatus of any kind, these men performed no experiments either for the benefit of their pupils or for their own intellectual advancement. (p. 142)

The sciences were not actually shunned or resisted by the existing academic elite. On the one hand, as I have mentioned, they were already an established, though minor, part of the curriculum by this time. They had been slowly admitted since as early as the 1740s and 1750s, when America's few colleges had undergone a degree of institutional reform in imitation of models provided by the universities of Edinburgh and Glasgow (this was especially true of William and Mary, which Jefferson had attended). In the early decades following the Revolution, when a number of states and the federal government began to underwrite natural resource surveys, schools such as Harvard, the University of Pennsylvania, and Columbia began to offer lectures in disciplines such as botany and natural history. Moreover, a number of the medical schools then attached to these colleges argued consistently for the addition of courses in "the new sciences," in the cause of updating instruction to "modern" levels.

Both before and after 1800 there was no organized or systematic resistance to science in American academia, either from classicists or from clerics. Even during the early years of the

Second Awakening, when religious fervor reentered the colleges, science was never attacked. Some, such as Yale's President, Timothy Dwight, opposed certain strains of Enlightenment and early romantic belief, whether deism or Rousseau's "natural man," and thus took a stand against "atheistic science." But their target was never science per se, but instead certain kinds of science. "Atheist science" could be countered by a "godly science" for most Americans. As exemplified by Benjamin Silliman, another early professor of chemistry (at Yale), technical knowledge and devout Protestantism melded easily in the idea that facts and causes in the material world were "the moral agencies of the Almighty," revealing "the character of the Creator as seen in his works" (in Brown, 1989, p. 104).

And yet, though neither scorned nor banished, science was not greeted by anything resembling a wholehearted embrace. Even as late as the 1820s, students received little more than a smattering exposure to scientific subjects, confined to a single semester of the senior year, compared with seven or eight semesters of study in Greek and Latin (see, e.g., Cremin, 1980, pp. 404–405). Ideals regarding the benefits of technical knowledge had not yet become a transformational force in America as they already had in France, Germany, and Scotland. They were the province of individuals, not institutions, in America, where the colleges remained too small, too provincial, and too obedient to the Latin–Greek tradition, with its emphasis on learning as the province of "eternal wisdom," to open their doors fully, or even half-way, to a group of new disciplines with real-world importance. Indeed, the ironic fact seems to be that science as a whole simply was not considered important enough by academics of the day for it to become the source of major controversy or consistent interest.

This uncertain status reflected a larger philosophical ambiguity with regard to the traditional "liberal arts." Because of its discourse of "fact," scientific thought could be adapted to the Academist view: Studying science did not conflict with the ultimate Academist goal of keeping the "liberal" mind free from specialized training of any kind, from "professional studies," and "mercantile, mechanical, or agricultural concerns" (see Guralnick, 1975a, p. 32). Yet throughout the revolutionary and immediate postrevolutionary period, science had been promoted among nonacademics—especially by politicians, but by scientists too—for its utilitarian powers. Whether the issue was the founding of an observatory or the promotion of a "native genius" such as the self-taught astronomer and mathematician David Rittenhouse, the advances promised nearly always came under the banner of "Abilities, speculative as well as practical, [that] would do Honour to any Country" (see Hindle, 1956, p. 168). No doubt the rhetoric that had succeeded in connecting science with revolutionary ideals contributed here as well. In one form or another, technical knowledge had been wedded to the notion of rendering nature into the hands of mankind. Thus its "purity" was at every step colored by the call to "utility." Accepted by the colleges for its devotion to "truth," science nonetheless warranted a peripheral place because of its "lower" connections to social existence.

This acceptance, moreover, took the form of absorption into, not simply addition to, the existing "liberal arts." Rather than being allowed styles of instruction fitted to their own kind of knowledge and their own methods, scientific disciplines were taught in the same way as nonscientific subjects. One learned physics in the identical manner that one learned Greek or rhetoric. One copied out for memorization a given set of lessons and lectures, prepared recitations for class, and underwent examinations demanding rote retrieval of the same. True, there were some demonstrations using some forms of laboratory equipment. Attention might be paid to various types of specimens, and perhaps on rare occasion an evening spent studying the stars. But these were all occasional exceptions to the basic methods of instruction: reading, memorization, formulaic repetition of given information, and increasing one's ability to express oneself properly, in oral or written language befitting a student of "higher pursuits." Anything resembling an interactive, experience-oriented type of teaching was almost entirely lacking.[2]

Moreover, students in science courses were often required to read, in untranslated form, various canonical Greek and Latin authors who had written on scientific subjects: Aristotle, Plato, Copernicus, Linnaeus, and so forth. This meant in effect (as Thomas Huxley once noted) that pupils had to be classicists first before they could even begin the study of technical knowledge. This method was a distinct holdover from the medieval and Renaissance university, and it remained the norm in Europe as well as in America. But in France and in Germany especially, such study was confined to the earlier years of college work; during the first half of the 19th century, it was increasingly succeeded by laboratory training, experimental procedure, and other aspects of what was already coming to be known as *Naturwissenschaft*, or "basic science." This changeover did not happen in the United States until much later. Though such terms as "pure" and "mixed" were used with regard to technical knowledge, science as a whole lacked any real independent status, either intellectually or practically. It commanded no separate loyalties and was not even taught on a regular basis in most institutions until the 1820s. Overall, there was simply no such thing as a "scientific education" in America, no courses whatsoever involving true experimental study, no granting of scientific degrees. Even Franklin's own Public Academy in Philadelphia, founded in 1756 to help further "the useful arts," had lost its engineering courses by 1811; everything scientific was banished to the senior year, where only a few token fields of study remained (astronomy, basic principles of chemistry and, of course—given the founder—electricity).

SCIENCE AND GENTLEMANLY PURSUITS

The fledgling state of the technical curriculum matched the larger place of science in the new Republic at this time. Provincial as it was, America did not yet have solid institutions to support and promote scientific work, academic or otherwise. The major exception to this was West Point, established in 1802 by President Jefferson, whose intent was to favor technical training over military training, to set up in effect an academy modeled on L'École Polytechnique in Paris which would then supply America with its own growing corps of engineers. Up until that time, nearly all technicians had come from abroad; indeed, nearly all of West Point's instructors were themselves

recent emigrés. Similarly, most science professors were also immigrants or had been trained in Europe. Few, if any, found the
opportunities for original research expansive. Indeed, no such
species as the professional scientist yet existed in America.
Instead, those Americans who referred to themselves as "men of
science" were largely self-taught amateurs, men of means whose
money and leisure enabled them to pursue independent learning.
They were necessarily autodidacts, dabblers and intellectual
tinkerers. Many viewed science as but one among many subjects
to pursue; few were exclusively devoted to it. To most, however,
science was a discipline that provided individual, intellectual,
and patriotic status all at once, a field of endeavor that helped to
shape the emerging image of the national gentleman-scholar.

Lacking a secure home, a solid organization, and a financial
base, American science during this period contributed little of a
fundamental, original nature to world knowledge. As Hindle
(1956, p. 383) notes, "Strikingly, no scientific research of the
time matched the importance of Franklin's colonial experiments
in electricity. Even in the realm of natural history where the
Americans did become more self-reliant . . . it is not clear that
Western science benefited as much as American pride."

The advancement of learning was not often the final aim of
these early "men of science." Rather, they were drawn by a range
of other benefits. They were interested in becoming worldly
moderns, in keeping up with their times and thereby representing "the American experiment" with dignity. Republican ideals
linking science to progress were always hovering in the background. But Franklin's concept of science-for-national-strength
was far less important than Jefferson's notion of "Nature's nation." Astronomy, geography, and natural history (including the
study of local plants, animals, insects, and rocks) were the
favored disciplines, for they fit a myth that portrayed "America"
as a vast living museum waiting to be explored. Indeed, Jefferson's own detailed program for the Lewis and Clark expedition
(1804–1806), carried out so successfully and famously, supplied
no small incentive for increased interest of this very type. That
this epic undertaking was able to merge adventure, nature,
science, and political objectives—in its effort to take possession,

by means of mapped and written observation, of the new territory included in the Louisiana Purchase (and beyond)—was not lost on the scattered community of natural historians at the time. It provided a grand exception to governmental policy in the form of a massive federal underwriting of science. It created a new, more rugged standard for bringing together "science" and "exploration," that carried through the rest of the century. It showed that a professional science might one day be possible—so long as it remained true to federal priorities.

But for the most part, American science in the first generation after 1800 remained a landscape of collectors, observers, catalogers, and conversationalists, profoundly distant from the great experimenters and theory builders of Europe. Prior to the 1820s and 1830s, when a frustrated generation of younger scholars, urged on by both shame and ambition, began a pilgrimage to universities and laboratories in Germany, merely a handful of Americans saw technical knowledge as a cause or mission for educational advancement at home. Again, a few outstanding exceptions existed, such as John Maclean at Princeton. Another was Benjamin Silliman, who briefly studied under Maclean before becoming Yale's first professor of chemistry in 1802 and going on to found the most successful scientific journal in the United States during the first 75 years of the Republic, the *American Journal of Science and Arts*, begun in 1819.

Yet, like Maclean, Silliman is instructive concerning circumstances that mitigated against what he himself came to stand for. He was not trained as a scientist; he had studied law. He wrote poetry, published political and satirical essays, gave speeches on religious subjects, and read widely in the arts, philosophy, and history. When offered the appointment as chemistry professor at Yale by President Dwight, a longtime friend, Silliman knew nothing of the subject whatsoever. After some study on his own, he journeyed to Philadelphia to attend lectures on chemistry and medicine, which he found inadequate. In the years 1805–1806 he traveled to England and Scotland, "ostensibly to purchase books for the college library but actually to learn something of the subjects he was supposed to teach" (Brown, 1989, p. xiii). While there, his interest shifted to include geology, and on his return to New Haven he began to offer lectures in mineralogy. Though gaining considerable fame both as an effective teacher and as a public speaker who promoted science education on the grounds of economic progress, national independence, and civic duty, his major impact was in planting the seeds for a professional scientific community that took more than 50 years to come to fruition. Indeed, only with America's full entry into the Industrial Revolution during the 1850s and 1860s, when many forms of technology became commonplace, did Silliman's dream of a home-grown scientific community, nourished by training at the college level, begin to appear.

In the meantime, Silliman sought to make the *American Journal of Science and Arts* an effective center of science education in the United States. His declared intention, spelled out in the inaugural issue, was to produce nothing less than the Frank-linian-Jeffersonian curriculum in magazine form:

> While Science will be cherished *for its own sake,* and with a due respect for its own *inherent* dignity; it will also be employed as the *handmaid to the Arts.* Its numerous applications to Agriculture . . . to our Manufactures, both mechanical and chemical; and to our Domestic Economy, will be carefully sought out and faithfully made.
>
> It is also within the design of this Journal to receive communication on Music, Sculpture, Engraving, Painting, and generally on the fine and liberal, as well as useful arts [including] Military and Civil Engineering, and the art of Navigation. (1819, pp. v–vi)

Silliman was highly conscious of his assumed role. "No future historian of American science will fail to commemorate this work," he wrote, "as our earliest *purely scientific* Journal, supported by *original American communications*" (1819, p. 3). The journal was intended to raise the level of study, learning, and publication to the level they had already achieved in Europe. At the same time, it was to be more: an embodiment of the ideal college, in which no learning of any kind was to be excluded, but where "the whole circle of physical science" formed the true core, with its infinite applicability "to human wants," its ability to "polish and benefit society," and finally, its revelation of the "supreme intelligence . . . harmony and beneficence of design in THE CREATOR" (p. 8).

Nonacademic Learning:
The Role of the Family

Though certainly one manifestation of scientific education during the time, the mostly self-taught gentleman-scholar was by no means the only one, or even the major one. The so-called common school, supported by local taxes, was already fairly widespread in New England, though much less so elsewhere, and typically offered rudimentary instruction in geography and natural history. Nowhere was enrollment or attendance required; students went when they were not needed in the fields, stores, or at home. Most schoolhouses were small and ill-kept structures, with single classes comprised of students from as young as 5 or 6 years to those in their later teens. Learning involved rote memorization and recitation by the whole class, and the main purpose of instruction was rudimentary literacy. Yet the names of many natural phenomena, such as the planets, trees, flowers, various animals, and so forth, were also included in the teaching process. "Science" was taught not as a distinct field of knowledge, but instead invisibly, as part of the larger catalogue of words and objects that had to be absorbed.

The more important learning in the sciences, meanwhile, took place outside the academic context altogether. Such learning was becoming important to the new middle class of prosperous farmers, storeowners, merchants, and others who sought to study and teach science in the home, in informal or semiformal groups, clubs, and associations, and in struggling and often failing natural history societies that depended wholly upon funding from their members. It is here, in the growing popular interest among this rising class of relatively affluent Americans, that the republican virtues were taken most seriously as a rationale for study.

To some extent, the nuclear family, which had itself come of age in the second half of the 18th century, remained a dominant educative force, embodied in this class. Outside the Latin schools—which charged fees and which were reserved for the wealthy few and nonwealthy but brilliant young men who were given scholarships in hopes that they would excel in such fields as the ministry, teaching, law, and medicine—practically all formal learning occurred at home or on the job. Sunday school was another frequent venue for learning, but one in which the Bible was the sole text and topic. In many middle-class homes the idea of apprenticeship was felt to be insufficient with regard to "education." More elevated notions of "culture," merged with the idea of raising good citizens and devout Christians, had become widespread. To varying degrees, both mothers and fathers took on the role of teachers, the transmitters of culture and cultural interest, with the intent of training the morals and raising the financial and social prospects of their children. The importance of home schooling is attested to by the high rates of literacy that existed, probably in the neighborhood of about 70% in 1800 (when less than 50% of children attended some form of school) and over 90% by 1840, as recorded by the national census (see Graff, 1987, pp. 248–257, 340–343).

Again, much of the science taught concerned astronomy and natural history. These were the subjects that could be experienced visually, and therefore made part of the discourse and experience of the home. As such, they could be personalized in their potential for "discovery" and made immediately available to the interested/fascinated amateur eye. They were the subjects, in short, that could be used to teach the precepts, moral and intellectual, of a loose and useful philosophy merging Baconian ideas with desires to comprehend the divine plan and its lessons of patriotic advantage.

In most homes of the period a kind of domestic curriculum was set up. Astronomy tended to be a "male" subject, taught by fathers to sons. Natural history, on the other hand, a far larger realm for potential seeing and learning, was handled mostly by mothers, who often took profound interest in the study of plants, flowers, animals, insects, local geography, and geology.[3] During this period, in fact, the role of the mother regarding domestic responsibilities was greatly broadened by new views on childrearing and family structure, and also by the renewal of religious sentiment, which mandated that the principal duty of wives was the raising and instructing of children. According to Cremin (1980, p. 65), "From an earlier role as aid and adjunct, she became the dominant figure of the family, creating with her strength, devotion, piety, and knowledge the ambience within which proper nurture could proceed."

It is interesting to find, then, that science was often a central part of this "nurturing." That a division of subjects developed along gender lines was perhaps predictable. But this division was less one of status or of viewed importance than of daily realities. When not required to work in the fields, the store, or elsewhere, women took responsibility for the home during the day and therefore tended to do the flower gardening, vegetable gardening, and other kinds of outdoor tasks that enabled them to develop an intimate familiarity with the natural environment around the house. Men who worked all day, especially in a town or urban setting, were essentially part-time inhabitants of this environment, visitors to its natural history. Women therefore tended to take charge of all things more immediate and of this earth; their space of knowledge was the space of life, of nature's daytime activity. Fathers who were farmers also lived in this universe and taught their sons and, to some degree, their daughters, about the natural history of the fields and surrounding woods. Yet for the increasing number of men who were employed in other ways, their science was most often that of the heavens, the night. The diurnal course of their working existence allowed them only the evenings for observation and learning. In spirit and in letter, their orbit of influence in the realm of knowledge was a more distant one.

In the home curriculum, therefore, a certain balance was sought: Both mothers and fathers worked to inculcate the values of careful perception, of drawing, recording, and discussing what could be learned from the natural world. These values, they believed, could be turned in their honesty and precision toward the world of men, the world of America. "Science" combined pragmatic and moral advantage, both of which could help build better citizens for a better Republic.

At the same time, as writers such as Kohlsted (1990) and C. M. Porter (1986) document, learning science within informal and semiformal groups and societies was very often connected to "American values." Kohlsted (p. 427) notes that "In the founding charters of natural history societies and learned academies, rhetoric about patriotism and presumed republic values in science and technology was commonplace." Since the Revolution, in fact, numerous small private academies had sprung up to serve the educational needs of local communities, still largely agrarian in composition and economy. Few of these "high schools" survived by midcentury, but it seems clear that together they represent the most far-reaching effort at educational reform on the institutional level during this period. Though hardly expensive by contemporary standards, they were still out of reach for most families of less than middle-class means (in addition to the tuition they charged, they also took male children away from work in fields, stores, offices, and so forth, and thus cut down on a family's overall earning power). Yet many of them enjoyed at least a brief period of success, especially in and around the urban areas of the northern states. Within them, science was often taught similarly and at similar levels of sophistication to what was offered at that time by the colleges. But differences did exist. The academies placed far less emphasis on fluency in Greek and Latin, and more emphasis on the practical aspects of technical knowledge, for example, in surveying, in agriculture, and in navigation. For these reasons, the colleges soon found themselves in direct competition with the academies for students.

SCIENCE AND AMERICA: NATURE AND VALUES

Thus, during the first quarter of the 19th century, the discourse about science among the new bourgeoisie was both more specific and more enthusiastic than the science discourse going on within the colleges. Where academia spoke highly of science yet treated it with relative disinterest, those with education, ambition, and means seem to have thought very much in the vein of Jefferson, when they journeyed, that is, beyond the limits of the "pious and polished gentleman." Study of science could promote discipline and patience, a sense of both reverence and understanding toward the natural world. It could provide a means to observe and to learn from the divine plan. It taught the value of duty to a higher, communal cause and enterprise, the worth of civility and loyalty to the Republic. Finally, it kept one up-to-date, helped make one an active part of "the new age," and therefore a better citizen.

Beneath this discourse, moreover, lay the still grander connection between science and pastoralism. Such a link should not be underemphasized; it permeated a great deal of thought and feeling about the new nation, which everyone knew to be filled with wilderness and the opportunities of exploration. The 18th-century slogan of a still-colonial America—"Westward the course of empire makes its way"—had not faded by any means but had only been renewed and invigorated by the news of Lewis and Clark's expedition and other explorations, by the rising notion of "the frontier," and by the sense of the unique and virginal nature that both enclosed and helped define the prospects of the country's future.

Images of the new nation as chosen by nature, gifted with a new world of natural things, abounded. Painters, poets, and journalists would soon take it up, just as science had already begun to do. Study of natural history translated, at this level, into discovering the physical body of the Republic. It also went hand in hand with the feeling that "Nature's nation" was also "God's nation," that the Almighty had applied his highest creative powers in forging the landscape of the new nation and there revealing, with bold and magnificent contours, the divine plan. As a young country, filled by wilderness, America lacked the glamor and power of European "culture." In its stead stood nature: wild, original, dynamic, sublime, and American. Guided by a romantic spirit blended with religious fervor, writers of the time often opposed this vital American nature directly against the Old World's crass, material civilization, where "despoiling hands" had transformed the land into a realm of trivial "garden landscapes." Yet if nature seemed to tug in one direction, and civilization another, these polarities nonetheless were brought back to unity in the forests of learning and knowledge. It was science, above all, that could link such anxieties, that could blend the religious, the patriotic, the intellectual, even at times the sentimental, into a single content for "America."

Looked at more closely, it could do even more. One sees that natural history as a republican venture also comprised an exegesis of another, more specific kind. It offered a needed and responsible acquaintanceship not only with the wonders of the surrounding environment, their unique domestic species and landforms, but also with the traits of America's national character written within. The idea that an American personality lay written in the rocks, waters, and organic life of the Republic, as something that would impose itself upon its citizens and help transform them into a race of heroes, was held by a majority of people at the time, in no small part due to the enormous reception given Samuel Stanhope Smith's *Essay on the Causes of the Variety of Complexion and Figure in the Human Species* (1787). In this book, it was expressed that human beings were descended from a single common ancestor and that subsequent differences among the various "races" were due to a mingling of environmental conditions (climate, scenery, latitude, etc.) and social conditions (the latter being often a subtle manifestation of the former; for a historical review of these and related ideas of environmentalism, endemic to much of the 18th and early 19th centuries, see Glacken, 1967, chap. 12).

Smith himself was a well-known writer and sermonizer on educational matters. As president of Princeton, he argued often for study of the sciences, whose learning he felt would teach discipline, patience, and piety. He saw human and national character as being mouldable on the one hand, yet best moulded in the direction that the natural environment intended.

Fundamentally, then, the most important educator in terms of the national future was the corpus materialis of the nation itself. Pastoral advantage went this far. An excellent and representative example of the standard images invoked, not merely by educators, journalists, and painters, but by scientists of the highest caliber, is again offered by Benjamin Silliman:

> With one glance, the eye will often survey extensive and luxuriant plains, covered with cattle, and rich in verdure; rivers flowing with a smooth and undisturbed surface, or roaring over rugged bottoms; hills crowned with orchards . . . valleys smiling with meadows and flowers . . . towns, villages and hamlets, indicative of rational life; and the immense ocean, lost at a distance beneath the incumbent sky. (in Brown, 1989, p. 77)

The mixture of qualities indicated in such a passage is itself as overwhelming as the landscape it seeks to describe. There is not only abundance, regularity, and "rational" life, but the vast play of scale and untamed energy as well. The power of the sublime and the beauty of the simple and bucolic are to be found everywhere in evidence, and promise a coming national ascendancy.

This type of pastoral advantage was also particularly visible in the reports of visiting or newly immigrant Europeans, especially those trained in natural history or geology. Witness, for instance, the following passage, taken from *Observations on the Geology of the United States of America* (1817), written by the Scotsman William Maclure:

> The peculiar structure of the continent of North America, by the extended continuity of the immense masses of rocks of the same formation or class . . . forces the observer's attention to the limits which separate the great and principal classes; on the tracing of which, he finds so much order and regularity, that the bare collection of the *facts* partake somewhat of the delusion of theory. (in Mather & Mason, 1970, p. 169)

MacLure was himself to play a role in the history of American education through his participation with Robert Owen in the New Harmony "experiment" during the 1820s (see Porter, 1986, for a detailed discussion of his career and ideas). His "scientific" inscription onto the land of such things as grandeur, symmetry, scale, the qualities of a reeling yet inspired consciousness, closely parallel the sentiment of Silliman quoted above. In a sense, the greatest teacher was nature itself, which stood between God on the one hand, and America on the other.

Such ideas, which attached the learning of things scientific to higher notions of piety and patriotism, did not wane in the following decades. The belief that nature was a type of holy manuscript given by God to his new chosen people was explored and given expression, in both word and paint, for nearly the whole of the 19th century. Well into the 1860s, for example, as Barbara Novak has noted (1980, pp. 7–8), the concept of the sublime—central to nearly all discussions of American landscape, wild or not had been largely transformed from an esthetic to a Christianized mark of the Deity resident in nature. Indeed the gradual fusion of esthetic and religious terms is an index of the appropriation of the landscape for religious and ultimately . . . nationalist purposes. Science, so prominent in the nineteenth-century consciousness, could hardly be left out either. Landscape, according to the mid-century critic James Jackson Jarves, was "the creation of the one God—his sensuous image and revelation, through the investigation of which by science or its representation by art men's hearts are lifted toward him" (*The Art-Idea*, 1864, p. 86).

Thus, in a sense, the sciences come full circle, back to the "liberal arts." Nature, in an established trope, becomes a canonical work created by the greatest of all great authors, and science is thus brought within the arms of literature in the realm of America-as-divine myth. That the contents of this myth, of the "reading" that those such as Jarves proposed, commonly yielded a message of universal meaning, of intimate relations with the divine reserved for a single people, was something endemic to the century as a whole. "Many of these projections on nature augmented the American's sense of his own unique nature, his unique opportunities, and could indeed foster a sense of destiny . . ." (Novak, 1980, p. 7). And this sense of destiny came to include and rely upon science, the social and moral engine of progress. Wherever nature went, after all, something of science was sure to follow. And though it sometimes stood in the wooly shadows, it often walked in a glaring light as well, and always through the landscape of an America growing ever more ripe with the ideals and the things that technical knowledge—divinely ordered—would bring.

PRACTICALIST CHALLENGES AND THE "YALE REPORT"

With regard to educational institutions of the early 19th century, however, one finds little to celebrate, much to confirm Henry Adams's "picturesque" diagnosis. Most of the societies, institutes, and academies formed during and up to this time for the study of science did not last. With a few exceptions (see note

1), patronage of scientific institutions for learning by the wealthy had not yet taken sufficient hold as a widely recognized means to advance the national "experiment." Those such as Samuel Stanhope Smith were among the few who ardently believed in the power of the "sublime sciences" (again, natural history most of all) to help shape the human personality and thus national character through the lessons of natural philosophy, by which he meant the Baconian trinity of observation, induction, and reasonable generalization (Smith, 1787, pp. 2–4). But his ideas were not strictly shared by the generation of *Monsieur-savants*, who came to power during this time. In their view, science was important for a mixture of other reasons, mostly having to do with personal/social/intellectual status, keeping up with (but not necessarily adding to) progress in the world, and patriotism with regard to America's image and honor (vis-à-vis Europe).

This general situation underwent a rapid change beginning in the 1820s, due above all to the advent of Jacksonian ideals. In the years following the War of 1812, resentment had continued to build against England and Europe as the seat of a "degenerate aristocracy." Older ideals that had originated in the colonial desire to duplicate the European society left behind, upper-class notions of "gentlemanly existence," forms of dress, speech, and behavior redolent of "genteel society," all became the objects of populist derision. In this climate of moral stand-taking and accusation—whose tone reflected the confusing mixture of sentiments, both celebratory and fearful, that came with loss of the old colonial social order—it was inevitable that the colleges would come under criticism. The role of collegiate education in the early Republic was small: No more than about 1% of the population ever attended, and most students did not remain long enough to receive a degree. But the criticisms were more than symbolic. As the site of highest learning, the college came to be viewed more and more as a possible realm for vocational training, both of an introductory and an advanced nature. A wave of Practicalist demand banged on the door of higher education, which was attacked for being out of touch with "the spirit of the age." During each year of the 1820s, it seemed, the fist of this demand grew tighter, harder; the knock louder, more urgent.

Specifically, it was a demand most often oriented toward upgrading the knowledge of farmers and merchants in the applied sciences. There were calls to help expand the economy, to ward off foreign competition, to advance opportunities for employment, and to equalize the benefits of "the national genius," which was now becoming evident in new technologies and methods of production. The colleges themselves were financially ailing, for several reasons. They suffered declining enrollments due to competition from local academies. Falling enrollments were compounded by the mounting accusations of "aristocratic" irrelevance, and more generally by the fundamental economic vulnerability of the colleges that often left them in a head-above-water circumstance in any case. Already, that is, and for some time, the colleges had been capitalist enterprises in the sense of both creating, and being entirely subservient to, a particular market demand system. In most cases funded by tuition more than endowment, they sought to attract students by charging a low scale of tuition and fees (as little as $35–50 per year), a system that ensured that even minor drops in enrollment would have significant deleterious effects. Being susceptible in this way, on a basis that their sense of mission and philosophy of knowledge despised, they were defensive about their social role. In effect, they had to protect their privilege by pretending it was above and beyond the Practicalist sensibility that made their institution possible.

The main problem was a fairly simple one. It can be perceived hovering within the various speeches and reports from college administrators of the time. Basically, the problem the colleges faced was to come up with a way to both accommodate (pacify) Jacksonian demands yet retain and protect the core identity of an Academist "college," an institution of restricted "higher" learning, loyal to "enduring truth"—which was now directly correlated with "freedom from subservience" to the mass public. The colleges, in short, knew that their own survival had become contingent on a willingness to change. They were willing to change, and they said as much: "We are decidely of the opinion that our present plan of education admits of improvement," began the famous Yale report of 1828. "We are aware that the system is imperfect: And we cherish the hope, that some of its defects may ere long be remedied." (Day & Kingsley, 1829, p. 5). Yet beyond such mollifying proclamations, change at the colleges was to occur only on their own terms. Such change was accomplished in several specific ways: (1) Greek and Latin, now often called "the dead languages" and perceived as symbols of the "aristocracy," were made optional; (2) the study of English was strengthened and expanded; (3) students were given more say in choosing their own course of study; and (4) science courses were given greater prominence—that is, their numbers increased, funding for scientific instruments and other supplies went up, and science became more firmly established in the curriculum. The physical sciences—astronomy, physics, geography, and chemistry above all—particularly benefited from the new collegiate accommodation with public demands.

None of these modifications were viewed as particularly serious or detrimental by the colleges. Indeed, they were easily offset by a new idea—the core curriculum—which henceforth would be necessary to "lay the foundation of a superior education." Students, that is, could select some courses according to their needs and interests, but only outside of a required nexus of subjects, whose intent was to provide the critical powers of intelligence their needed form and destiny.

The Yale report formed the document of warrant for this belief, and was perhaps the most influential position paper on higher education written prior to the Civil War. Published under the title *Original Papers in Relation to a Course of Liberal Education* in Silliman's *American Journal of Science and Arts* (Day & Kingsley, 1829), and, due to its reception, as a separate volume the following year, the document was a direct response to Practicalist attacks on Yale, both by the public and by students. A great deal of apparent soul- and mind-searching went on in the several years leading up to the writing of the document, done by Yale's president Jeremiah Day, and a well-known classical scholar, James K. Kingsley. The result, as historians have pointed out, was a manifesto of conservative higher learning that has framed educational arguments in the United States ever since.

As put by Day, "The two great points to be gained in intellectual culture, are the *discipline* and the *furniture* of the mind; expanding its powers, and storing it with knowledge"

(1829, p. 301). The mind, then, was pictured as a sort of home requiring organization, control, and taste. More, its natural structure was comprised of a fixed series of "mental faculties," which, in the rationalist (Cartesian) psychology of the time, formed the interior nucleus to the body's exterior "house":

> As the bodily frame is brought to its highest perfection, not by one simple and uniform motion, but by a variety of exercises; so the mental faculties are expanded, and invigorated, and adapted to each other with the different departments of science. . . . From the pure mathematics, [the student] learns the art of demonstrable reasoning. In attending to the physical sciences, he becomes familiar with facts, with the process of induction, and the varieties of probable evidence. In ancient literature, he finds some of the most finished models of taste; by English reading he learns the powers of the language in which he is to speak and write. By logic and mental philosophy, he is taught the art of thinking; by rhetoric and oratory, the art of speaking. By frequent exercise on written composition, he acquires copiousness and accuracy of expression. (pp. 301–303)

Day does not mention the "faculties" by name. Instead, he describes them in terms of their endpoint, productive capabilities. In fact, these faculties were presumed to be the same in the child as in the adult. They were fixed, unchanging entities of the human mind, including such things as memory, inference, judgment, and language; what they required was "exercise" for their expansion, not growth or maturation. In its fully developed version the relevant psychology listed a total of 37 separate faculties (which included such things as aggressiveness, veneration, benevolence, and so on, in addition to those emphasized by pedagogy); it served as the basis for the theory of Phrenology, which sought to literalize the presumably fixed nature of these faculties, their mental-organ character, by assuming their topographic influence on the head itself. In any case, those faculties selected for emphasis by educators of the day bore a rather striking resemblance to the medieval curriculum itself, from which they were no doubt derived. But their static quality, the immobile character of mind they presumed, matched perfectly as a rationale the idea of a curriculum consisting of an equally fixed series of canonical works and timeless truths.

The college, as it then stood, thus became a kind of "natural" institution. The individual's intellect defined a kind of core-type series of categories; and the college, in its very organization and through its established teaching method, precisely mirrored mental structure itself. Behind Day's defense, in other words, was the remarkable claim that college is excellent for the mind because, in a sense, it *is* the mind—in its highest possible development (thus the idea of the college as being architecturally contained and removed, and its general partitioning into so-called faculties today, take on particular historical resonance).

Note too that the balance of this institutional intelligence, meanwhile, is heavily weighted on the side of the humanities. Mathematics and the sciences are mentioned first, but just to get them out of the way (Day's "different departments of science" obviously uses the word in its broadest sense, that is, as knowledge generally). To accommodate Practicalist demand, the curriculum is strategically, if gently, turned upside down; what follows are the more profound faculties, those that allow the student to think, speak, and write in broad and essentialist terms, of which the sciences then appear as subsets.

This naturalizing of the college had another element as well. This was the element of "parental superintendence": "The parental character of college government requires that the students should be so collected together, as to constitute one family" (Day & Kingsley, 1829, p. 308). Thus, as surrogate father (though never stated as such, this is the clear intent), the college enacts a curriculum, together with a method of teaching, that continues the protected growth of the student. Colleges are in the business of advanced child-rearing; performed according to the lessons of nature, derived from the very structure of the mind.

Kliebard (1992, p. 7) has characterized the Yale report as "the most comprehensive account of the conventional wisdom on the undergraduate curriculum . . . and the most spirited defense in its time of the humanistic curriculum against the onslaught of modernity." But one should consider that it also layed out and helped solidify the fundamental epistemology and moral scheme for the "core" idea itself—and this idea, in modified form, has remained central to the Academist position down to the present day. Indeed, one finds it everywhere in evidence during the 1980s and early 1990s (characteristically, the 1989 NEH document *50 Hours: A Core Curriculum for College Students* cites the Yale report approvingly). The notion of establishing and teaching a core curriculum as a means to define (and determine) "what it is an educated person should know" remains one of the most influential ideas in American higher education. In fact, the faculties psychology has never died out completely with regard to the curriculum. It has reappeared over the last 150 years, like a ghost in chains, haunting every claim made that the study of a particular discipline will somehow lead inevitably to generic forms of mental superiority, for example, "studying a second language gives us greater mastery over our own speech, helps us shape our thoughts with greater precision and our expressions with greater eloquence" (NEH, 1987, p. 13).

In any case, the new core of the late 1820s and 1830s was an interim success. It reflected an effort to both expand the curriculum in terms of offerings yet stabilize and finalize it with regard to working concepts of "higher learning." The largest expansion took place in the realm of the sciences: It was here that the more significant accommodation was made to Jacksonian demands for "relevance." But with few exceptions the new openness to science did not involve a similar openness to real practical training, in the form of engineering or agricultural courses, for example. These were "sent down" to the high schools, academies, agricultural seminaries, or else to the strictly professional schools; in any case demoted to a lower, more "commercial" order of study. The purpose of the college was to impart a deeper "substance," that is "that various and general knowledge, which will improve and elevate, and adorn any occupation" (Day & Kingsley, 1829, p. 349). Thus did the colleges separate themselves from, and raise themselves above, their immediate competitors. Thus did they rationalize their role in a democracy as providing the trappings of "an aristocracy for everyone." Thus, too, however, did they show themselves really interested in attracting only a more rarified type of student (though as many of them as possible), drawn to "ennobling the faculties" more than employing them.

The sciences now added to the curriculum—geology, botany, meteorology, natural history, and so on—were taught in a "purist" manner, severed from real-world problems, as the abode of new knowledge about the physical universe. One did not study "science" as the Practicalist phenomenon it was often coming to be in society at large, involved with the new idea of "technology" (itself made popular in a series of lectures given at Harvard in the late 1820s by the physician, Jacob Bigelow, and published the same year as the Yale Report). One studied science as part of the new core, as a modern addition to an older system of removed and elevated truth.

Soon, within a decade or so, by the mid-1840s, teaching methods did begin to change. Illustrated lectures, experiments (sometimes assisted by students), and even field trips (in "outdoor" disciplines such as geology or botany), had become commonplace. Inside the contained space of academic experience, a more participatory realm was opening up, beginning at the periphery first. But again, this change should not be misconstrued. It did not involve any important shift of power relations within the academy. Rather, all was offered in the name of initiating the student into the ways of science—science, that is, as an Academist idea and a growing academic profession.

The Age of Jackson and After, Part 1: Popular Imagery and Public Reformers

The period between 1830 and 1860 was one of the most critical in American history with regard to concepts concerning science, education, and their interrelationships. It was a time of great public celebration over the forms and practical effects of technology, yet also of suspicion about "pure" and theoretical science for its aroma of elitism. At the same time, scientists were busy forging the basis for a true professional community, a community that came into being within the colleges and was based on the idea of research and "higher truth." Finally, the Jacksonian era was also a period when beliefs about education reminiscent of those held by Jefferson and Franklin were revived, and hardened into public dogmas. But in the process of this hardening, these beliefs were deeply and irrevocably altered, transformed into cultural movements of sensibility that have remained in place down to the present, and into ideas that employed concepts and images derived from science in a variety of fascinating ways.

Given the fundamental importance of all these developments, which occurred in but a few short decades and were some of the most central features of the change to a fully modern society in America, I feel it necessary to provide some idea of their scope and complexity. For it is in terms of their adaptation that the subsequent historical patterns acquire meaning. I will therefore pause on this era to look at some of the ideas, the institutions, the individuals, and the images that helped finally make both education and science intrepidly "American."

POPULAR IMAGERY

The age of Jackson was marked by the first great wave of science popularization among the public in the United States. Within Jacksonian rhetoric, in fact, technical invention and "the common genius" were interchangeable. The signs of material advance, after all, lay everywhere—the proof that America's time had arrived. Between 1820 and 1840 industrial development had begun to advance enormously in the United States, and at an unprecedented pace. Railroads, steam power, the cotton gin, and a dozen other innovations and new technologies within a few decades helped to expand the types of industry and varieties of employment for both the middle and lower classes beyond all prediction. Part of this expansion involved more complex, rationalized forms of labor and management. To build and run the new railroads, the mill towns, the factories, and great southern plantations, and to bring new technologies to market, generally required imposing schemes of coherent organization on both men and the materials they employed: "Systems" became one of the watchwords of the day. Whether under this term, or that of "invention," science and technology became unifying sources of imagery for progress, itself an important new slogan for the sense of maturing national destiny. Science and technology promised a more glamorous future—profitable, secure, exciting—to communities, interest groups, and individuals who all sought a stake in the new age of "improvement."

But one should be clear about what type of science was being favored. This was not theoretical science, not an intellectual knowledge about such things as heat, light, motion, or force. Nor was it Jeffersonian science or Franklinian "usefulness," in the sense of being either universal truth or an educative knowledge to make the individual more capable in life. The "science" of favor at this time was instead eminently practical: For most people, "invention" was the source of grand new productive power and possible affluence. Steam engines, railroads, the power loom—such innovations were the shared testimony to national material wealth, to America the brilliant and the bountiful. "Pure" science (a term already in use by the 1820s) was perceived to express a kind of aristocratic disinterest in public needs. In the antebellum period, no belief yet existed in any necessary or probable link between fundamental scientific inquiry and eventual utilitarian result. What existed instead was "ingrained suspicion of the man of useless knowledge" (Daniels, 1968, p. 47).

With regard to popular imagery, then, the primal power of technical knowledge lay in a particular realm. What was this imagery, and how widespread did it become? Authors such as Leo Marx (1964) and John Kasson (1976) have characterized the rhetoric of the time as being one of "the technological sublime" (see, e.g., Marx, 1964, p. 195). It was a rhetoric of utter intoxication and fascination before the spectacle of mechanized power, what Eli Whitney called the "proof of ingenuity put to use." Technology—and by implication, science—became grandly, even wildly aestheticized. A thousand benefits and assurances of every imaginable type, economic at first, but then broadening out to encompass all regions of human endeavor and need, from the spiritual to the domestic, came to be invested in machines. For most popular writers and much of their audience, advances in the "mechanical arts" were viewed as signs of America's distinctive claim to greatness. Whether treated as evidence or as allegory, the machine seemed to offer a kind of culmination in which ancient mythic prophecies were being fulfilled through the hard work of republican morale and ingenuity. Articles, speeches, and books of the 1830s, 1840s, and 1850s regularly appeared with such titles as *The Moral Influence of Steam* or *The Gods of Invention*. The machine held out the promise not only of national unity, but of social equality. Jacksonian writers who otherwise took the colleges to task for their outdated and impractical classicism nonetheless often cannibalized the most revered contents of the classical canonical texts to celebrate the glory of the machine, employing, for example, the myths of Prometheus, Orpheus, and Herakles, the *Iliad*, the works of Milton and Shakespeare, and a hundred other monuments of academic piety to describe the exalted powers recently released upon the earth by the "mechanical arts." More generally, ambient faith in the machine sometimes verged on the rapturous. No less an orator than Daniel Webster put it thus:

> It is an extraordinary era in which we live. It is altogether new. The world has seen nothing like it before. I will not pretend, no one can pretend, to discern the end; but everybody knows that the age is remarkable for scientific research . . . and perhaps more remarkable still for the application of this scientific research to the pursuits of life. The ancients saw nothing like it. The moderns have seen nothing like it till the present generation. (in Marx, 1964, p. 214)

In such an atmosphere, it was inevitable that inventor such as Robert Fulton would be dubbed "American Archimedes" and would become, along with Benjamin Franklin, the object of cultlike worship. The endowing of machines with symbolic force inevitably endowed their creators with the same. Both were elements of the new, material national destiny.

On the other hand, the technological sublime also had its broad-scale pedagogic attributes. This means, in effect, that certain aspects of the mechanistic were adopted as models, teachers in a sense, for various areas of work and thought. Ideas of "technique" and "scientific law," for example, emerged like communal themata in a wide range of fields and arts. Landscape, portrait, and genre painters of the time wrote openly of the need to develop technical powers, both of observation and depiction: as Asher B. Durand put it, ". . . every *truthful* study of near and simple objects will qualify [the artist] for the more difficult and complex; it is only thus [he] can learn to read the great book of Nature, to . . . transcribe from its pages" (in Novak, 1980, p. 9). Such artists sought to employ an exact craft in their promotion of "the American landscape as an effective substitute for a missing national tradition" (p. 20).

Whether sublime or sentimental, the emotional response aimed at—a re-creation on canvas of the most profound nature experiences—could be achieved only by blending together the mechanical skill of the trained painter/draftsman with a near-scientific eye able to perceive God's plan, as written in the details of the American scene. For the most famous and successful painters of the day—those such as Thomas Cole, Durand,

Albert Bierstadt, and Karl Bodmer—to paint this American scene was to create a kind of visual pedagogy that set moral man in the unique grandeur and spirituality of American reality.

In literature too didactic purposes could be achieved via technique. Often, this involved the idea of authors developing their skill through formulaic imitation of a few given, standard themes that rendered the novel or story into a moral textbook. Writers who treated the epic struggles and individual lives of the revolutionary era, the early presidents, tales of the wilderness, and the like, were frequently praised for their "functional artistry" or "efficient narrative." Science more directly entered into the literary domain through ideas of nature, for example, in the person of James Fennimore Cooper's "natural man," Natty Bumppo, developed in the five-novel series of Leatherstocking Tales. Bumppo, in effect, was a kind of walking moral: Selfless, "taught by life," he revealed the stout essence of American national character in his simultaneous rejection of savagery and effeteness and his deep knowledge—moral and systematic—of the frontier. He possessed, as Cooper once described it, the true science of the wilderness. His understanding of the forest, the Indians, the streams, seasons, and weather was full of technical detail. He could read nature's signs, decipher its every movement, translate these meanings into instant action, pure decisiveness. The landscape had been "the laboratory of his maturation," and he himself had become "its living instrument."

Within educational thought itself, however, the machine and its derived instrumentalism were an even more evident and clearly stated model. Here, several things might be said. First, a specific and important change was instituted in private secondary classrooms (with the exception of the "pauper" schools, no public system of education yet existed). Beginning in the 1830s a new rationalizing attitude toward education, based on ideas of "system" and "function," gained strength both among educators and the public. One result was an entire, sweeping changeover in classroom structure—from mixed-age schoolrooms, in which children sat and performed their lessons individually (often in neglect), to age-segregated classes in which "entire classrooms of children passed as one through a common set of exercises: rising and falling in unison, reciting together in choral sing-song, or

conning their common lesson with their bodies locked in precisely the same posture" (Rogers, 1985, p. 122). The rationale
for this change drew its sustenance from the mills and factories of
the Northeast, where the benefits of well-controlled, orderly
mass behavior had been proven, and where the concept of
replaceable workers—like replaceable machine parts—had been
pioneered.

The metaphor, in fact, is irresistible: The process of learning, still fundamentally based on the psychology of faculties,
seems to have shifted from notions of "exercise" to notions of
technological regularity, to forms of mechanical obedience and
inculcation. And, indeed, the rhetoric of the day surrounding
education is full of such mechanistic images. Both within and
outside the common school movement, schools were compared
to clocks, to millworks, to engines, to "smooth and well-run
machines" in general. Mechanical imagery, moreover, extended
to other elements as well, arguing for standardization.

"We know not how society can be aided more than by the
formation of a body of wise and efficient educators," wrote
William Ellery Channing in his well-known *Remarks on Education* published in 1833 (see Welter, 1971, p. 45). Or, as expressed in the *Common School Journal* of 1840,

> When the system and methods of a good normal institution ha[ve]
> begun to produce their direct effects on schools, and their results are
> carried to towns and villages . . . and their children are made subjects
> of these experiments, how will they rise in the estimate of the public!
> . . . By means of the cultivated and well directed intelligence of the
> instructors, operating with the machinery of good methods, the
> phenomena of society will be explained to the young, its framework
> and processes in some degree made known, the principles taught by
> which men and women ought to be guided in life. (in Welter, 1971,
> p. 75)

Essays and speeches often waxed rhapsodic about education
as the "generous unfolding of the whole spiritual being." Yet
when it came time to discuss the actual activities of teaching and
learning, the individual disappeared into a sea of industrial
imagery, centered on efficiency, on method, on the yielding of a
useful and well-formed product. Education was to take place by
means of the endless repeating of given material, through the

creating of "mental habit," through the training of reflexive discipline. In this way, the child would learn not merely the lesson itself (which would ingrain itself on his or her memory like a template), but such important life habits as punctuality, social union, cooperation, and restraint of personal impulse. Thus school was reimagined from an instrumental institution into an instrument in its own right, a utopian device that would help guarantee an America that would itself come to resemble a well-oiled invention, orderly, reliable, and infinitely productive. This image was to recur later on in the century, with merciless insistence. The school-as-factory model is one of the most pervasive and durable images in the history of the United States and, predictably no doubt, was itself an important part of the larger change wrought by the coming of industrial society. Finally, as I mentioned above, this had its echo in other educational institutions. As Kasson (1976, p. 63) writes, "Public schools, parks and gardens, art galleries and museums, Sunday schools and tract societies all represented attempts to extend the sphere of republican instruction in the principles of social order and virtue to the maximum number of citizens; to counteract the turbulence and corruption of American life by improving the social environment and establishing monitors over it." Museums, in fact, went from being simple *Kunstkammer*—rambling private collections of souvenir objects and curios, stuffed into a few rooms—to more open, public spaces that sought to guide the viewer in a regulatory fashion through a hierarchy of classifications. Moreover, the building of penitentiaries for criminals, houses for the poor, asylums for the insane, and orphanages for the homeless—all these had their educational intentions as well. Physically designed on the model of factories or mills, these institutions were, in a manner of speaking, schools for the sake of those whom society had failed to (or could not) educate to its proper ways of life. Regarding the penitentiary, for example, the central belief was that "parents who sent their children into the society without a rigorous training in discipline and obedience would find them someday in the prison" (Rothman, 1971, p. 70). In a very literal sense, then, the prison or orphanage or almshouse was a direct replacement for the educational environment that had perhaps been denied the inmate when he (or she) was a child. It was the "corrective school" of adulthood.

THE COMMON SCHOOL MOVEMENT AND SCIENCE

Industrialization brought its fair measure of disturbing change to the established order of what had been largely rural, agricultural America. The new forms of large-scale production and wage labor disrupted older social relationships based on local markets, personal contact, and small-scale, often family businesses. Increasing social and economic mobility destabilized many communities, while the cities absorbed large new populations, growing rapidly without adequate support institutions and public works. The first great episode of immigration from Europe took place during this time, resulting in separatist "cultural villages" and sources of cheap labor.

The sense of impending chaos and uncertainty was palpable, both in urban and in rural areas, and helped give a strong impetus to the swoon over technology and mechanistic social models promising order, predictability, and monitorial control. As I suggested above, such an environment placed education high on the list of priorities as a means for restoring control, in which humanitarian, economic, moral, and nationalist aims could all be merged. Jacksonian populism, in fact, seized on the idea as a kind of collision point of moral reference, a subject on which to base arguments for the defeat of elitism, the uplift of general opportunity, but also the inculcation of proper habits, attitudes, and obedience for the sake of a better Republic. Between about 1825 and 1840 these goals were all brought together in the form of the common school movement, which grew up around a group of reformers whose energy and dedication were as tireless as their moralistic fervor. Inspired in part by early socialist thinkers such as Robert Owen, and following Jefferson's lead with regard to learning as essential to liberty, these school reformers included Josiah Holbrook, Stephen Simpson, Henry Barnard, James Carter, and Horace Mann, all of whom, in one form or another, battled for a public system of education that would be open to all for the purposes of individual and civic betterment and for the creation of future generations founded on a "sharing of values." In fighting for such a "system" of public education, these men often employed ideas and metaphors that either called upon science directly or that drew upon it as a form of authority:

How can it be a marvel, that wealth practises oppression, when it holds as its allies, all the riches of knowledge, and the exterior semblances of virtue and truth? Moving in the high orbit of science, government and laws; ordaining justice and morality after their own images, how shall we ever counteract the principles of vassalage that now prevail, unless we procure EDUCATION for our offspring, and diffuse SCIENCE among our brethen? . . . Knowledge is power, in respect to the procurement of equity to the great mass of the sons of labour. (Simpson, in Vassar, 1965, p. 193)

For socialist reformers such as Simpson, science and education were largely synonymous. The great enslaving force within society was the maldistribution of knowledge and the know-how it could provide to further both individual and civic betterment. The notion that an enlightened public would best resist all attempts at tyranny, and would one day act to melt away all social inequities, was prevalent at the time, in Europe as well as the United States. In America, the difference often lay in the continuing influence of Scottish philosophy, which insisted upon the essential liberating power of scientific knowledge in particular. Interpretations of this power and how to put it to work, however, varied a good deal among reformers. In the end, science came to be used as a theme in a movement that merged liberal and conservative causes—that created a truly open, public education, yet one decidedly fixated on forms of order, standardization, and hierarchical contents.

Behind much of the common school rhetoric lay a sense that the family had somehow failed and that society was therefore in trouble. Inculcating moral virtue—the acknowledged goal that would provide the solid bedrock to an informed and patriotic citizenry—gained a certain strident tone; it could no longer be left to the individual. The "state of society" was responsible for the evils in evidence, for class divisions that destroyed the self-respect and required internal authority of families at the lower end of the economic spectrum. "Never will wisdom preside . . . until Common Schools . . . shall create a more far-seeing intelligence and a purer morality than has ever existed among communities of men," wrote Horace Mann (in Cremin, 1957, p. 7), and in this he echoed many other thinkers of his day. Society therefore had to take back nature from the home. Society, in the form of an institution that would provide free education to all, had to act as a counter-weight to the "impure" morality that had weakened this home and the community around it. Thus, in curious concert with the Yale Report of 1828, did common school reformers such as Mann mix in a guardian role, full of "parental superintendence," with the more political notion of universal learning in the cause of freedom. Built in the form of a proper "system," on the basis of the principles of "moral science," society could erect an educational institution that, by virtue of its parental function, would make human beings more natural and therefore more free.

The first great institutional success of the common school movement—one that continued for decades thereafter—was the American lyceum movement, inspired by Josiah Holbrook in the late 1820s. From the beginning, this scheme took as its main goal the dissemination of scientific knowledge, which Holbrook viewed as best achieved through a unique blending of formal and informal schooling, systemic in organization yet fully encouraging of individual self-advancement through "mutual instruction." The lyceum therefore placed more emphasis on the individual, *sensu stricto*, than did the programs of other reformers. It was not, however, without its hierarchical, parental side as well. Holbrook had studied at Yale under Benjamin Silliman, well-known chemist, mineralogist, promoter of science education, and editor of the *American Journal of Science and Arts.* Silliman's dignified and unshakeable enthusiasm had been poured into Holbrook, who soon after launched himself as a traveling lecturer on scientific subjects. His experiences on tour revealed to him, he said, a hunger for learning widespread among the citizenry. In October, 1826 he announced his own plan for a system of American lyceums, whose constitution, published two years later in the *American Journal of Education* (Vol. 3, p. 503), began with the following two articles:

> Article 1. The objects of the lyceum are the improvement of its members in useful knowledge, and the advancement of popular education, by introducing uniformity and improvements in common schools, by becoming auxiliary to a board of education.
>
> Article 2. To effect these objects, they will procure a cabinet, consisting of books, apparatus for illustrating the sciences, and a

collection of minerals, and will hold meetings for discussions, dissertations, illustrating the sciences, or other exercises which shall be thought expedient. (Knight & Hall, 1951, p. 145)

Thus aimed at science and "the useful arts," the lyceum was to be built upon a recognized need for both "uniformity" and open "discussions" among "members." Supported by minimal entry fees and conducted as a voluntary and mutual teaching enterprise for interested individuals, the lyceum nonetheless sought to become "auxiliary to a board of education." It was not really a school in the institutional sense, but rather a collective, a "town meeting of the mind," as Carl Bode (1956) has aptly called it; and yet it sought a degree of institutional status, a measure of the privilege and respect already granted to existing formal educational bodies. The basic internal organization was similarly loose yet conventional, including a hierarchy of elected officers, involving fixed meetings and lectures, yet deeply dependent on revolving contributions by individual members.

This unique balancing of "parental" and individualistic elements had an unprecedented success. By the late 1830s, a mere decade after Holbrook's proclamation, there were nearly four thousand lyceum scattered across the country, comprising a total membership of more than one hundred thousand people. Speakers as well known as Ralph Waldo Emerson and Daniel Webster had become regular lecturers on the lyceum circuit. Many lyceums continued strong into the 1860s, to the brink of the Civil War and after. Taken as a whole, the movement was the first true form of popular education in the country and by far the most significant impetus to public education early on—both as example and as a forum for the spread of ideas and attitudes. In this alone, it was enormously powerful in its influence on the whole of the 19th century.

For the most part, individual lyceums were a varied lot. They formed in cities, in towns, in fishing villages, in farming areas, and at factory centers. The topics for lectures and discussions often changed in concert with the kind of community the lyceum served. Some met weekly, some monthly, many irregularly, others constantly. Part of their attraction lay in the fact that their mixture of formalism and spontaneity allowed for individual interest at many levels of commitment (from mere gaping to intensive efforts at self-improvement). Often enough, Holbrook's overriding focus on things scientific and technological held sway. Nature came to have an immediate and concrete meaning: The science of plants, rocks, stars, animals, and machines was something that all Americans could share and trade among themselves. Science did not merely lead or encourage democracy; within the bounds of these organizations, it *was* democracy, embodied in objects and visibilities of the native land. Study progressed by means of a combination of assigned readings, discussions, experiments, study of minerals, and, above all, through lectures, the most public form of exposure. Field trips were sometimes held, excursions for rock and plant collecting became favorite activities, telescopes and other instruments were made use of, funds permitting. In more rural areas, agricultural science and medicine tended to dominate the lyceum curriculum. Subjects were often chosen by mutual vote and thus took a popular cast, on the basis of recognized social or immediately practical aims, or else on fashionable interest. Bode (1956) provides examples of lecture topics for a given season that include such things as: the Practical Man, the Life of Mohammed, the Honey Bee, Popular Education, Geology, the Sun, the Poet of Natural History, the Legal Rights of Women, Instinct, and the Capacity of the Human Mind for Culture and Improvement. People wanted to become familiar with many things: the nature of the Republic, developments in astronomy or medicine, the "lessons of history," new fertilizers or methods of planting, recent inventions, human nature. As was characteristic of the Jacksonian age, there was often a clear emphasis on use-value (e.g., as expressed by the title of one of the most powerful and well attended of the lyceums, the Boston Society for the Diffusion of Useful Knowledge). Academist notions of higher learning for the elite were roundly taken to task—not the least through the idea of science as knowledge destined "to raise the standard of common education" and thus to advance the level of the common man (Holbrook, in Knight & Hall, 1951, p. 146). The movement in its heyday involved a substantial portion of the new middle class across the country. As a whole it both reflected and advanced the public image of "science" as a source of moral, intellectual, and patriotic virtue.

Yet, however Practicalist in spirit, the lyceum had decided Academist leanings. Even the word itself, denoting the garden of covered walks where Aristotle instructed students in his philosophy, bespoke a certain classical orientation. As a title, moreover, it already had wide international currency in Europe before its adoption by Holbrook. The Paris Lycée, for example, founded in 1786, was a formal institute where famous professors came to lecture on literacy and scientific matters to a discerning (upperclass) public. Only a few years afterward, Napoleon made it the lower-case term for his state-centralized secondary school system, the "common school" of France. In Britain, meanwhile, "lyceum" was a term used to designate many highbred literary groups, and could be found chiseled above the entrances of the imposing, often ornate buildings erected to house them. Holbrook's borrowing, then, was itself imitative in part, something that carried with it certain status suggestions of Old World learning and tradition. But more than this, the American lyceum practiced a deceptive informality. Being firmly wedded to the lecture platform above all, it became a center for the most authoritarian of educational experiences. Lyceums helped launch the careers of many famous speakers; indeed, they brought the lecture, as a literary and money-making form, to new heights of style, popularity, and influence. Men such as Emerson could make a good living from it, and, in their decades of itinerancy, were thereby urged to give the didactic tone of the time a declamatory grandeur. Lyceum "stars" replaced political figures and religious leaders as the focus of public attention and the embodiment of eloquence. They exposed "the people" to the "great minds and ideas" of the day—ideas whose presumed importance, of course, the lecturers themselves (if skillful) were felt to incarnate through the power of their words, the modulation of their voices, and the aristocratic bearing of their persons. The lecturers thus formed a kind of circulating service elite for populist consumption. Being at once eminent and popular, they were eminent popularizers of the Academist ideal. Beyond everything else, they expressed and gave warrant to the attitude that, while always potentially available to the common man, higher learning was still the province of higher individuals.[1]

HORACE MANN AND RALPH WALDO EMERSON: IDEAS OF EDUCATION AND SCIENCE

Thus, during a period of unmitigated revery toward technology, science, and education, certain Academist attitudes and beliefs remained unshaken, even entrenched. Certainly, those of the intellectual elite who had been schooled in such beliefs were not about to abandon them overnight, whatever their hopes and plans for reform. More important still was the related feeling that Jacksonian populism had its unruly, excessive side, that too much power and self-determination, unguided by properly elevated faculties, would lead directly to the "rule of King Mob." Education with its presumed capacity to shape and form the common man in a higher image (one with specific attributes, to be sure) was thus envisioned as a means to preserve social order, at the same time that it was presumed to promise an end to inequality and tyranny. Such were the confusions and utilities of the age, that Jeffersonian rhetoric could remain virtually intact, yet now be wedded to more conservative desires for control and predictability.

Indeed, the prevalence of these attitudes among reformers, especially with regard to science, is nowhere better revealed than through a brief look at two of the period's most prominent mandarins of influence. No discussion of education during this early, critical period in American history can, in fact, avoid dealing with these two figures, who, in many ways, through their contrasting careers of pragmatic action and philosophic removal, reveal how certain tenets of traditional Academism composed a fundamental orthodoxy during this time, uniting those of otherwise disparate backgrounds and ambitions.

Even in terms of basic educational experience, no more striking opposites could be found than Horace Mann and Ralph Waldo Emerson. While Mann came from a small farm in rural Massachusetts and was largely self-taught before entering college at age 20, Emerson was born into a prominent Boston family of intellectual ministers, and was formally educated at Boston Latin School before entering Harvard at age 14. Their subsequent careers were even more dissimilar. Mann strove to become a great builder of institutions, both as a legislator and as the first secretary of the Massachusetts State Board of Education, where he made public education his mission, eventually succeeding in his efforts to found a tax-supported free and open school system that soon became the model for the entire nation, an achievement that earned him the title of "the father of American education." Emerson, on the other hand, personified the man who stood apart from practical affairs and political strife. Rather than a builder or believer in institutional systems, he cast lofty doubt and Romantic distrust upon them at every turn. To Mann's missionary zeal—he wrote at the end of his own life, "Be ashamed to die until you have won some victory for humanity" (in Cremin, 1957, p. 27)—Emerson might well have responded, "He is great who is what he is from nature, and who never reminds us of others" (1983, p. 617).

On the surface, Horace Mann certainly seems a mix of Jacksonian and Jeffersonian sensibilities. As a legislator in the Massachusetts House of Representatives from 1827 to 1837, he campaigned relentlessly for all manner of humanitarian causes: He tried to help the blind, the mentally ill, and new immigrants, and he supported the abolition of slavery. His most essential plea, however, after accepting the low-paying position of education secretary, was a clear extension of Thomas Jefferson's fundamental Enlightenment belief that a truly free society, with the possibility of economic and moral advancement for all, was impossible without a free, universal system of public education. The emphasis, however, had changed: The perfection of mankind was no longer the focus. Instead, a new Practicalist note had been introduced. "An educated people is a more industrious and productive people," Mann wrote. "Knowledge and abundance sustain to each other the relation of cause and effect" (in Cremin, 1957, p. 61). Learning must now serve society, in other words, and society, in turn, must pledge itself to "the education of all its youth, up to such a point as will save them from poverty and vice, and prepare them for the adequate performance of

their social and civil duties" (p. 77). Mann saw universal schooling as a means to unify and harmonize American society. As
Lawrence Cremin (1957, p. 8) puts it, his desire "was to use
education to fashion a new American character out of a maze of
conflicting cultural traditions." He envisioned a universal collection of secondary schools that would be a true "system," "bind-
[ing] all into a scientific whole" (p. 45). It would do this,
moreover, by inculcating social responsibility via such things as
discipline, attentiveness, and acceptance of authority—moral
virtues that would both "uplift the mind" and render it capable of
"productive toil."

In many ways Mann anticipated contemporary attitudes.
His basic values were purely middle-class ones, and his success
might well have been predicated on the rise of this group during
the middle part of the century (see, e.g., Kaestle, 1983). He
argued that schooling should be extended to girls; that corporeal
punishment was harmful and counterproductive; that rote learning was oxymoronic; that schoolrooms should be decent public
buildings; and that teaching, "the most difficult of arts, and the
profoundest of all science," (in Cremin, 1957, p. 21) must be
seen as a form of expertise that required aptitude, a desire to do
the best job, and proper training. He felt, too, that education
was the great "experiment" which would abolish all distinctions
among children related to class or background, "all the miseries
that spring from the coexistence . . . of enormous wealth and
squalid want" (p. 88). He thus shared the liberal's view that good
schooling could overcome and remedy every social evil, even
those most private and familial.

But these things should not be used to obscure the central
purpose at hand: Education should be used to cultivate "moral
virtue," and to instill good behaviors and "salubrious attitudes,"
to save children "from poverty and vice." School was to be a
kind of protectorate, in that it was where the "nobler faculties"
could be raised to "dominion" over "the intemperate passions."
This concept of the school was itself based on Mann's view of
human nature, which he very often framed in scientific or
technological images. Despite talk of "purity" and "innocence,"
he favored a basic distrust of the child—not as the site of
Original Sin but as an undetermined beginning that could fall to

either side of the line. "Men are cast-iron; but children are wax,"
he wrote in his diary, immediately after accepting the position of
state education secretary (in Mann, 1865, p. 5). If "the preserva-
tion of order, together with the proper despatch of business"
were ever disrupted, "the school will be temporarily converted
into a promiscuous rabble." This did not argue for "military
formality"; yet it did portray the school as a place often "rescued
out of chaos," therefore always demanding of "government and
discipline" (in Cremin, 1957, pp. 49–50). On the other hand,
"system compacts labor" (p. 49). One must therefore cast the
future republic by patterned craft, in a technically consistent
mold. Every child must be "trained to self-government," in this
efficient image, for this is what freedom really means, this
beneficent yet designed consensus (p. 57). Mann thus appealed
to the mixture of bourgeois feeling that saw in the common
school, at one and the same time, the instrument by which the
state took a family's children away from it but also constituted a
means to better discipline and advance their children in a more
complicated, mechanized world. Such a view of human nature
also had its immediate political demands, in terms of education:

> If republican institutions do wake up unexampled energies in the
> whole mass of a people, and give them implements of unexampled
> power wherewith to work out their will, then these same institu-
> tions ought also to confer upon that people unexampled wisdom
> and rectitude. . . . In a word, we must not add to the impulsive,
> without also adding to the regulating force. (in Mayfield, 1982, p.
> 168)

The larger parental function of the public school could be
conceived in technological terms. It would "regulate"—
discipline, contain, redirect—certain potentially disruptive
"energies" set loose (among the childlike "mass") by democratic
self-determination itself. It would put the brake on "impulse"; its
fundamental work was to replace nature with culture. "A republican form of government, without the proper education of its
people, must be, on a vast scale, what a mad-house, without
superintendent or keepers would be" (in Cremin, 1957, p. 90).

Such was why, too, "beyond all other devices of human
origin, education is the great equalizer of the conditions of

men—the balance-wheel of the social machinery" (p. 87). For, in the end, "It may be an easy thing to make a Republic; but it is a very laborious thing to make Republicans" (p. 92).

Education, therefore, was the ultimate productive engine for a better America, for a society that itself resembled an organic machine. "Our means of education are the grand machinery by which the 'raw material' of human nature can be worked up into inventors and discoverers . . . into the founders of benevolent institutions . . ." (in Cremin, 1957, p. 79). Education was the power center where society was built, calibrated, and set to working.

One should note, however, that the productive and regulatory function of learning would come not from some new and reformed program of study but instead from the standard curriculum, which, in the Massachusetts system Mann implemented, continued to put primary emphasis on language arts and arithmetic. Under his purview, classroom teaching emphasized the "exercise" of mental "faculties" above all else: simple computation; the reading of a few standard texts; the copying out of assigned compositions; ritual oral recitation of religious material, civic codes, and the like. Language, above all, was the "indispensable condition of our existence as rational beings" (in Cremin, 1957, p. 34); thus "reading becomes the noblest instrument of wisdom" (p. 43). Everything for the student came directly and entirely from the teacher. Everything for the teacher, came from a wholly conventional, Academist course of training. Though promoted and argued on the basis of scientific images, this basic curriculum did not, therefore, emphasize the learning of science itself. Like many of his contemporaries, Mann stressed "moral science" rather than physical or natural science. He departed strongly from Jefferson in believing that true virtue was something that could come almost exclusively from the study of humanistic subjects—the lessons of great thoughts. Shakespeare and the Scriptures would help instill the highest values most directly. And while the sciences might serve this aim in other ways, by their example of Baconian patience, observation, and reasoning, they were not central to Mann's scheme for education. On the plains of order and ethics, they were never the equal of the "great examples of human virtue" embodied in great writers, in biblical lore, in the "liberal arts" generally.

Emerson, at first, presents a somewhat different picture, an oratorial and even iconic rather than legislative presence in the history of 19th-century education. Certainly, his canonical ubiquity in studies written today makes him a necessary object of study. Emerson, in fact, has often been called Horace Mann's opposite, in that he favored a type of radical individualism that seemed to deny or drench with skepticism any establishment, orthodox program for learning.

Yet this perspective is only partly true. His stress on the heroic self and its capabilities for uplift actually led him to favor the same type of basic educational content as that promoted by Mann and other contemporaries with Academist leanings. Emerson's tendency to skid into idolatory of the magnificent (mostly literary) individual forced him into a dilemma; reading his work today, we see him caught between a Jeffersonian belief in the everyman and a more standard conservative, even elegaic preference for great "leaders of Nature," men who would both personify the national spirit and inspire their fellow citizens by imaginative work and example.

Emerson's most direct statement of his educational views was given early on in "The American Scholar" (1837), often called "America's intellectual Declaration of Independence." An address delivered to the highly elite Phi Beta Kappa Society, it sets the foundation for higher learning squarely on the shoulders of "those who do," scholars and researchers in other words, while nonetheless casting doubt on the value of learning itself. Higher education, like that of the schoolroom, Emerson portrays as a required passage of rules and routines that amend the deficits of nature. Though not welcoming to greatness, its parochialism is nonetheless often necessary as a lesson for going beyond this kind of imposed learning. What types of knowledge, then, are of most worth? The author clearly prefers a more-or-less standard Academist program, a reading of the great thinkers in search of inspiration. Part of this is his promotion of the scholar over the professor—for it is the former who must be a "university of knowledges," who carries the grandeur and responsibility so often misattributed to his institution.[2] Teaching the classics is a means by which students are exposed to exemplary wisdom and selfhood, and such teaching (the suggestion runs) is best carried out only by those who are the living embodiments of such wisdom, or at least who live its atmosphere on a daily basis. In his moral capacity, the scholar becomes the container and the medium of past greatness. Great teachers are great thinkers and workers first; their teaching flows from other abilities. Given such sentiments, it can hardly be a surprise that the one educational movement Emerson backed with a real personal commitment was the lyceum (which also supplied him with his main livelihood for more than a decade). Though he did speak of the need for common schools, so much of his thought stood against institution making that such support must have sounded sometimes contradictory.[3]

In one thing, however, Emerson provided a direct parallel to Horace Mann. While he constantly employed the natural sciences as a reservoir of rhetorical material, and saw science itself as capable of great social and cultural benefit, he did not make science education, as a subject in its own right, central to his ideas. This does not mean that he ignored or avoided such education. On the contrary, it was part of his larger vision for mental experience, for learning that would ennoble the individual and expand the ethical substance of the nation. But it was a minor part, at best, and subordinate to the literary model.

In a speech given to the Mercantile Library Association of Boston, entitled "The Young American" (1844), Emerson declared:

> The bountiful continent is ours, state on state, and territory on territory. . . . The task of surveying, planting, and building upon this immense trace, requires an education and a sentiment commensurate thereto. . . . The arts of engineering and of architecture [should be] studied; scientific agriculture is an object of growing attention; the mineral riches are explored; limestone, coal, slate, and iron. . . . We cannot look on the freedom of this country, in connexion with its youth, without a presentiment that here shall laws and institutions exist on some scale of proportion to the majesty of nature. (1983, pp. 214, 217)

This is Emerson at the rare peak of his direct advocacy for education in the sciences. It is an advocacy for a certain type of science—Franklin's "useful arts"—that becomes a means for defining "America" as a content of grand fertility, of bounty for the alleviation of inequality and suffering. Technical training, like technology itself, can achieve its ethical truth by serving to redistribute material welfare. Yet Emerson's attitude toward technology, more generally, was a mixed one. As summarized by Kasson (1976, p. 134), though he often granted "the benefits of technology, he insisted that the imaginative and moral life which was the ultimate justification of a republic could never be reached by mechanical means." America, in other words, must have a still higher content than that of technological knowledge, which tended to employ and train the imagination more than to elevate it to its noblest level.

That could be done by another type of science, a more pure and intellectual science, the birthright of canonical leaders. "The great man makes the great thing," Emerson writes in "The American Scholar"; "Linnaeus makes botany the most alluring of studies, and wins it from the farmer and the herb-woman; Davy, chemistry; Cuvier, fossils" (in Emerson, 1983, pp. 65–66). In "Self-Reliance," he states this another way: "Every new mind is a new classification. If it prove a mind of uncommon activity and power, a Locke, a Lavoisier, a Hutton, a Bentham, a Fourier, it imposes its classification on other men . . . and so to the number of the objects it touches and brings within reach of the pupil" (in Emerson, 1983, pp. 276–277). The highest form of learning in science thus has a religious quality to it; students are less discoverers than disciples, their minds bent to the classification of the master, a prophet of knowledge. Science achieves its exalted form in the superior figure, who invests his discipline and all that he touches with superior life. The scientist is thus the literary hero in different clothes.

Emerson's mythic yet traditional Academism is evident in another way with regard to science. The sciences form a topic to which he pays much creative lip service by way of metaphorical borrowing, but which otherwise gain little admittance to the curriculum of his writing as the subjects of actual discussion. At times—for example, in his introduction to *Representative Men* (1850), entitled "Uses of Great Men"—his narrative literally teems with images derived from various scientific disciplines (e.g., "As plants convert the minerals into food for animals, so each man converts some raw material in nature to human use"; in Emerson, 1983, p. 618). And yet there is much irony in this:

Not one of the individuals the author goes on to discuss, after all, is a scientist (they include Plato, Swedenborg, Montaigne, Shakespeare, Napoleon, and Goethe). Moreover, not one of his many essays is devoted to science, nor is it adopted as a subject in any extended portion of those such as "Culture," "Nature," or "Intellect."

If Emerson is a lens, we must look into him from both ends of the telescope. Though he places science at a certain distance, his use of it as a source of imagery magnifies other, more popular notions of the day, those that go beyond what we saw in Mann. On the one hand, he laments the frequent dryness of scientific knowledge ("Our botany is all names, not powers . . . science has put the man into a bottle," he writes in "Beauty"; see Emerson, 1983, pp. 1099, 1101). Yet such knowledge, when infused by higher sentience, becomes part of mankind's own creations of greatness. This becomes clear in Emerson's larger view of science, which he links with certain beneficent visions of progress, prevalent at the time, to be sure, yet to which he gives his own magnificent stamp: "The motive of science [is] the extension of man, on all sides into Nature, till his hands should touch the stars . . . and, through his sympathy, heaven and earth should talk with him" (from the essay "Beauty," in *The Conduct of Life,* 1860; see Emerson, 1983, p. 1100). Or again, "The destiny of organized nature is amelioration, and who can tell its limits? It is for man to tame the chaos; on every side, whilst he lives, to scatter the seeds of science and of song, that climate, corn, animals, men, may be milder, and the germs of love and benefit may be multiplied" (from "Uses of Great Men," in Emerson, 1983, p. 632).

By this time in American history certain fundamental terms of Jeffersonian belief regarding science and its teaching had been reversed. Instead of extending the nature within man, and thereby aiming at perfectibility, the high moral power of science was now to extend the mind and spirit "into Nature," there to achieve new forces of order, organization, and communion, all for mitigating the harsher and more primitive demands of existence. No doubt, it was too late to believe in human perfectability; but improvement through methods that would "tame the chaos" of the world—the natural cosmos and human society both—was a necessary and entirely ambient stipulation for the idea of raising "America" and the individual to the level of their own greater destinies. Emerson and Mann came at this goal from different directions. For Mann and other common school reformers, the drive was to fashion the individual in society's better image. For Emerson, it was society that should be forged in the image of the individual, as a sympathy, often mystical, between self and the world.

Both men, however, believed in the virtues of order acquired through noble aspirations. Both felt that the most "representative men" whose example might serve to inspire such hopes were literary or political, not scientific. The times demanded that one speak of science and invention, creatively, perhaps opportunistically, at the least with some degree of concern or respect, as a popular touchstone. As a subject and a model, the "scientific" proved a basis for the founding of the American lyceum, for thinking of public education in terms of "system," even for justifying a rote type of learning. But none of this guaranteed that science itself would be central to most programs or philosophies of learning. In an era of Practicalist ambition, Academism continued to rule as the final curriculum standard. To create a better generation of minds for the future, the new order would best be achieved by a certain fundamental obedience to the old, to preferences for humanist studies and the moral advantages they were assumed to inevitably bring.

The Age of Jackson and After, Part 2: Professionalization of Science and the Role of Higher Education

For the first 25 or 30 years of the 19th century the image of science was emerging as an important connection of power—moral, political, intellectual, economic—that both crossed and divided classes, interests, and institutions. In the scheme of higher education, as David Noble (1977, p. 20) has written,

> The colleges were firmly in the hands of the classicists and the clerics, and there was considerable academic disdain . . . for the teaching of the "useful arts." Technical education in the United States, therefore, developed in struggle with the[se] colleges, both inside and outside of them. One form of this development was the gradual growth of [scientific] studies within the classical colleges, resulting from the reorientation of natural philosophy toward the empirical, experimental . . . search for truth and from the pressures of some scientists and powerful industrialists for practical instruction; the other was the rise of technical colleges and institutes . . . in response to the demands of internal improvement projects like canal-building, railroads, manufactures, and, eventually, science-based industry.

Yet this type of summary, though perhaps accurate in broad outline, conveys little of the drama with which change would come about after 1840, when new forms of power became associated with science, and more directly, technology—power symbolic, power mythic, and power professional. Neither can such a précis suggest how reforms at the university level were but a part of a much larger environment of cultural revision, in an age of social upheaval and confusion marked by the search for new forms of order. The era spanning 1830–1860, after all, was the time when America thrust itself forward into industrial modernism. And part of this modernism involved the Americanization of science, as idea, as image, and as profession. What changed, in short, was both the popular and the elite images of science— the styles of popularization not only among the general populace, as discussed in the previous chapter, but among scientists themselves, who now sought new styles of heroism to attach to "the scientific." And what resulted, by the 1860s, simultaneous with the fervid celebration of native technology, was the ascent of "research" as the hallmark of true scientific work, the beginnings of a true scientific community, self-aware and activist, whose final home was in the colleges and whose work was therefore closely allied with teaching, and with the creation of future generations of scientists.

Science as Academic Profession and as Professional Image Making

The change from amateurism to professionalism within science was a result of several trends. One was the accelerated advance in technical knowledge itself during this time, which helped alter the broader character of this knowledge and its production. Mathematics had come to play a more important and central part, and experimentation had spread to almost every discipline. Science was evolving into a sphere of work and thought that self-consciously differed from other intellectual areas, that sought its own ground. College reforms of the 1820s and early 1830s, meanwhile, which had introduced more science into the curriculum, and the increasing influence of such popular teachers as Benjamin Silliman at Yale, had helped considerably to swell the ranks of trained scientists, most of whom had been taught to believe that their professional training made them superior to the "self-made man of science" of previous years. These factors, from about 1830 onward, led to the founding of many new specialized journals, each the visible sign of a new kingdom of expertise. The estimates provided by Daniels (1968, pp. 231–232), for example, show the total number of scientific journals increasing from 27 in 1820 to 50 by 1835 and to more than 90 by 1850. American science was now much less the work of isolated, self-taught individuals. It had become "a collective enterprise . . . [employing] labor, capital, and management" (Bruce, 1987, p. 4). Both a consumer and producer of social materials, science had entered a historical stage where the organizational realities associated with modern professionalism began to dominate.

This professionalization took several forms (for a general discussion, see Reingold, 1991, pp. 24–53). One involved the founding of scientific societies and associations endowed with sufficient funding and membership to endure. In contrast to previous decades, the new societies were no longer local, confined to a single city or centered on a single museum, and consisting of only a few dozen members. In 1846 the American Association for the Advancement of Science (AAAS) was established (in direct imitation, one might note, of the British version, set up in 1831) with an initial membership of 461; in less than a decade, this grew to over a thousand. Representatives at its annual meetings came from every major college and academy in the United States, from scores of smaller scientific groups, and also included a large number of persons simply interested in the sciences—for example, journalists, writers, and public officials. Early on, these meetings were more general in scope than they were scientific in any strict professional sense. But this circumstance itself became a subject of much debate within the association: By the late 1850s leadership was controlled by academic scientists who entertained only topics of "true scientific seriousness" (see, e.g., Kohlstedt, 1976; Bruce, 1987 pp. 256–260).

Even before this time another source of professionalism had arisen. This involved the creation of separate technical colleges, entirely distinct from the existing Academist system. Among these were Renssealaer Polytechnic Institute (founded in 1824, reorganized in 1851), the Ohio Mechanics College (1828), Cooper Union (1859), and a range of local institutes focused especially on the teaching of scientific agriculture. These schools provided people with a range of backgrounds, from urban areas, towns and rural districts alike, with the type of practical and theoretical training the older colleges refused to provide. Aimed at producing well-educated engineers and scientists for utilitarian or entrepreneurial work, these new colleges were commonly underwritten by gifts from wealthy donors. Most were industrialists and merchants responding both to a perceived lack of home-grown technical talent and to the felt absence of such training in their own background. Such men were the agents of Practicalism on an important new scale. Their motives were complex, often involving a blend of self-interest, philanthropic intentions, and patriotic fervor. Peter Cooper, for example, founded Cooper Union (full title: Peter Cooper Union for the Advancement of Science and Art) as a free night school for working-class adults, on the model of L'École Polytechnique in Paris:

> What interested me most deeply was the fact that hundreds of young men were there from all parts of France, living on a bare crust of bread a day to get the benefit of those lectures. Feeling then, as I always have, my own want of education, and more especially my want of scientific knowledge as applicable to the various callings in which I had been engaged, it was this want of my own . . . that led me, in deep sympathy for those whom I knew would be subject to the same wants and inconveniences . . . to provide an institution where a course of instruction would be open and free to all who felt a want of scientific knowledge, as applicable to any of the useful purposes of life. (1864; in Vassar, 1965, vol. 2, p. 192)

Cooper and other new patrons of practical science believed ardently that training in technical matters far more than in the "higher arts" alone was absolutely necessary for building, strengthening, securing, and advancing the United States, the source of their wealth, privilege, and opportunity. Beyond this, these plutocrats shared a faith in science as a teacher of moral lessons (patience, cooperation, diligence), and this faith, merged with their populist hopes for a more benevolent and orderly society, made them willing participants in the general celebration of technology and, inevitably, turned them toward trying to effect change through education. Thus, often enough, though they themselves had sometimes passed through the existing college system, such men felt slighted and angered by its granitic stand against the kinds of utilitarian knowledge for which they had come to stand. Founding new schools was therefore their way of both expanding that system and gaining revenge against it.

As for scientists themselves, one should note that by the 1830s they had already become employees of the universities, and therefore constituted a de facto profession. They were educators; this was their first calling. Though research would come to occupy an ever greater portion of their time and energy, especially from the 1840s onward, they were nonetheless centered in the colleges and usually saw teaching as an essential part of their mission to expand science through the conversion and encouragement of their students. Indeed, the curriculum reforms of the 1820s and 1830s, though limited in overall extent, had nonetheless laid the groundwork for a true scientific professoriat in the United States. Even before 1840 academia had become the main professional home of American science. It could hardly be a surprise, therefore, that a majority of the bright and ambitious science students would look for ways to remain within this setting, seeing the college as their inevitable residence, or that their professors woud try and aid this process by opening new space for them, hiring them as assistants, part-time lecturers, and so on until new teaching positions could be created. (Advancing the careers of individual students at this time was frequently equated with the advancement of American science.) Actively, though often quietly, but in the long run effectively, the professors of the early 1800s came to comprise a relatively coherent force for change merely by pursuing their own careers as teachers. From a total of about 25 in 1800, the science professoriat in antebellum American colleges grew to around 60 by 1828, during the first round of reform, and thereafter burgeoned to over 300 by 1850 (Guralnick, 1975a, p. 142). This growth was linked as well to a greater interest in science on the part of college presidents, who worried about the survival of their institutions during a period of renewed public attack for "irrelevance" in the late 1830s, 1840s, and 1850s, when techno-enthusiasm was at its height, and when, etched against this background, the traditional colleges again appeared "irrelevant." Boards of trustees at all the older schools were gradually persuaded to expand the curriculum in science, therefore opening the doors to the hiring of more scientists: During the 1830s geology became standard at nearly every school; during the 1840s, mineralogy and physics were added; in the 1850s, biology came into its own; and by 1860 courses in botany, zoology, agricultural chemistry, and other fields were widespread.

Educator and, increasingly, researcher too, the new science professional was also sometimes a consultant. Low pay and secondary faculty status, together with the reigning academic disdain for practical work, helped encourage an increasing number of scientists to seek or accept temporary commissions in industry and government in order to profit, economically and symbolically, from the utilitarian applications of their knowledge. Geologists agreed to organize and lead a range of state and national surveys; mineralogists performed assays and other analyses for mining companies; physicists found employment in the redesign of engines; and chemists, most active of all, were hired by textile merchants to work on improving dyes and bleaches, by the military to devise new or improved explosives, and by museums and wealthy individuals to perform laboratory analyses on different substances. Thus, to some degree, the modern career of the academic scientist had been discovered, in relatively full if nascent form, even by this time. Being a professor-scientist already bore the marks of several occupations at once, drawn together by the (presumably) original duty of teaching.

That this scientist had chosen a particular institutional home meant, then, that his professionalism was profoundly tied to the demands and controls endemic to that institution. These included: the establishing of reputations; the importance of publication; the framing of intellectual standards and certification; the need to locate sources of outside support; the required yet tenuous balancing of educational and other responsibilities; and, last but certainly not least, the adoption of an Academist rhetoric (based on "higher truth") that still largely defined the link between knowledge and teaching on which the college itself was based. These demands led scientists to accept the established self-image of the academic life, and to adopt this into their evolving professional discourse.

On the larger cultural scene, therefore, scientists found themselves walking a narrow and wobbly fence, one that shook and leaned between public representation and insider image. While the public continued its swoon over technology and could be sold "science" if utilitarian arguments were used, the colleges in their classicism preferred only a purist "science," unsullied by such practical concerns. Two sciences, then, were involved, one populist and technological, the other elitist and theoretical. As a result, a range of arguments, fit to different interest groups, were devised. As Daniels (1968, p. 50) remarks:

> The government was told that national honor demanded that science be given support and the public was told of the vast practical benefits that science had to offer. The conservative fear of social disorder was effectively exploited by pointing out the quietistic [moral discipline] function of science . . . , and the more liberal were offered the prospect of bringing social "elegancies" within the reach of all classes. Once this rather comprehensive argument was developed it is not suprising that the possibilities of science so quickly captured the imagination of Americans, and the verbal—if not yet the pecuni-ary—support of most public officials.

Such a "comprehensive argument," one should note, was nearly identical to that advanced by Horace Mann in his program for public education as a whole. It was not that scientists borrowed or begged such tactics from social reformers such as Mann. The reality seems to have been that these types of rationales—in their seizure of claims for both pragmatic and epistemological progress—defined a shared reservoir of argument for all institution builders in the Jacksonian era, an essential touchstone for claims making, for justifying new constructions of authority.

In particular, carving an Academist niche for science in the late antebellum period meant, first of all, establishing the idea of the scientist-as-researcher. This, in turn, proved possible be-cause of changing developments in the worldview and specific labor of science. Much depended, for example, on the recognized place and power of the laboratory as a site for manipulating and purifying nature, and thereby for setting the scientist up as the generator of new truths. The laboratory was the site where natural phenomena, often invisible and uncontrolled in the outside world, were forced to show themselves, becoming accessible to the eye, the hand, and the pen. But still more, it was the place where scientists could claim new portions of nature for their own, the stage where they took increasing possession of the world itself, where the Earth and everything on it became increasingly scientific. This, then, helped aid the liberation of science from philosophy. "Experimentation" and "analysis" now replaced "speculation" and "logic." "Results" and "hypotheses" now shoved aside the former importance of "facts" and "specimens." The rise of laboratory work in the United States occurred in direct response to the great and renowned advances made in this area in Europe: Visits to German laboratories, especially to that of Justus von Liebig in Giessen, where biochemistry and scientific agriculture were born, had convinced American scientists even by the early 1830s of their own evident and embarassing inferiority, and therefore of the need to expand opportunities for research in the name of national pride (indeed, of all the Practicalist rationale that had been used to promote science in earlier decades, only this goal of promoting national honor penetrated deeply enough into the professional self-image of scientists to remain central throughout the 19th century). American scientists of the 1830s, 1840s, and 1850s were heavily dependent on books from Europe, findings from Europe, theories from Europe, often equipment from Europe, and, worst of all, training in Europe. And they were acutely aware too of how one-sided was the dependence. Europe was their model and their framed antagonist. The professionalism and public support for science witnessed there, by increasing numbers of American intellectual visitors, urged them to speak of their own work in elevated tones of discovery, and about the "higher essence of science" in the work of promoting national goals.

Between 1820 and 1850 there thus occurred a definitive shift away from interest in amateur work, with its Baconian empiricism, and toward the theoretical, the quantitative, the conceptual. Most scientists had come to feel in any case that Bacon's ideas concerning the discovery of truth-through-induction had proved empty and fruitless: Masses of specimens and observations had been collected, yielding no laws or principles as promised. New forms of study and analysis were required. Research—as a mixture of lab testing, hypothesis making, and article writing—became the professional norm. By the 1840s leading scientists such as Joseph Henry, the country's foremost physicist, were calling the Baconians and all autodidacts "charlatans": "Our newspapers are filled with puffs of Quackery and every man who can burn phosphorous in oxygen and exhibit a few experiments to a class of young ladies is called a man of science" (see Miller, 1970, p. 7). The amateur, in short, had become something more than a mere fake; he was a seducer and corrupter of "innocent minds."

On the other hand, it was indeed a time of great theory building in Europe. Even before Darwin and Helmholz, Lyell had advanced uniformitarianism in geology; Dalton had postulated atomic composition; physics had been altered by Laplace's work; Theodore Schwann had pioneered the cell theory in biology; and Sir Humphry Davy, Lorenzo Avogadro, J. L. Gay-Lussac, and J. J. Berzelius, among others, were inventing modern chemistry. Leaders of the American scientific community often spoke of the "explosion" of European science taking place at the time. The notion of scientific work as a critical, self-correcting search for "fundamental principles"—a more secular notion than the older belief in uncovering material revelations concerning the divine plan—gained great force as a means for American scientists to join with and even compete with their European teachers, to obey yet appropriate for their own use the Academist rhetoric of higher wisdom, and to infuse everything with a quality of destiny that united professional and national prestige.

This did not mean that the new breed of scientist despised or ignored technology, for example for its mercenary aims. On the contrary, those such as Joseph Henry or Alexander Dallas Bache, two of the most influential professionals of the mid-19th century, often admitted the progress and advantage brought by invention. But they did so on a particular basis. Technological

enterprise, Joseph Henry wrote, "at this time so zealously pros-
ecut[ed] in all parts of the union, emphatically teach[es] us how
mathematical and mechanical knowledge may be applied not
only to the necessities of ordinary life, but to increase the power
and promote the wealth of a nation" (in Reingold, 1991, p.
130). Technology was thus the outcome—indeed, an inseparable benefit—from deeper understanding of mathematical and
scientific principles. Henry, and other like-minded researchers,
were not (as Reingold aptly points out) pure advocates for pure
science, insincere with regard to utilitarian goals. Rather, they
helped establish, in their day, an important link of faith through
their belief that "every technological field embodied elements of
the theoretical" (Reingold, 1991, p. 132). While it is too simple
to interpret this belief in terms of "pure" and "applied" science
(as if, e.g., the latter were a necessary effect of the former), it
nonetheless, during that time, effectively set up an order of
significance, with research (pure or applied) as the most fundamental kind of scientific work.

Regarding education in particular, the new brand of researchers, a number of whom banded loosely together in a group
called the Lazzaroni (named after a collection of Renaissance
Florentine beggars), tended to emphasize the need to train more
scientists, to expand the facilities for professional work within
the colleges. In was time, in their view, "to make science more
respected at home . . . to increase the facilities of scientific
investigators and the inducements to scientific labours" (Henry,
in Miller, 1970, p. 9). Thus enlightening or informing the
public was not foremost on their agenda, except where it might
aid the larger goal of "a bounty for research," as expressed by
Bache, leader of the Lazzaroni. What the public needed instead
was to be illumined about the needs of science, American
science above all. "Is it the diffusion of science, or the
encouragement of research that American science requires,"
Bache asked rhetorically in an address to Congress (1844) on the
subject of "The Condition of Science in Europe and the United
States." "Is it sympathy and kindly communication of which we
have the most need?. . . Or opportunities, means and appliances
for research? Do we need talkers or workers?" (in Miller, 1970,
p. 5). From this perspective, the new science of professional

researchers was one that the public should behold mainly as spectators, fascinated tourists in the land of "the magnificent quest." It was not a garden to be planted for required pragmatic yield.[1]

Such, certainly, was the decided if ironic message of the enormous popularizing campaign launched during the 1850s and 1860s, with Louis Agassiz at its head. Conceived largely by the Lazzaroni (Bache in particular), this campaign adopted the aim of raising the level of interest and support for American science by tapping as chief spokesman a worldly European who, himself, had been conveniently converted to the "infinite opportunities" that the new republic seemed to offer (as Agassiz himself put it). Thus did Emerson's "great man" descend to the level of celebrity-hood. With a position at Harvard secured, Agassiz embarked on a series of public lectures that succeeded in garnering unprece-dented sums of money for scientific institutions. According to Edward Lurie (1960, pp. 214–248) in the seven years between 1853 and 1860, Agassiz gave hundreds of lectures to packed halls in cities all across the country, amassing almost $600,000 for a variety of projects including museums, natural history col-lections, publications, and technical drawings. The level of private sponsorship of science probably doubled or even tripled during this time. Such success was ensured in no small measure by Agassiz's own foreignness: Most wealthly, educated Amer-icans, after all, were themselves schooled in the Academist tradition, and still viewed Europe as the true home of culture and intellectual magnificence. By means of his Germanic accent, his dramatic gestures, his stately bearing and proud face, Agassiz played to the reigning stereotype of the continental grand hom-me, providing an appearance at once imposing and familiar. His charm and influence, his ability to promote science through his Europeanisms, made him as much a willing foil of the age as a part-time leader of the public mind. In his many, often flam-boyant, yet always skillful and entertaining speeches, he held forth on the "uniqueness" of American natural history, offering the Old World equivalent to the "great ideas" of science. And he did more. For by succeeding as he did, he also gave evidence on a deeper level about what the new researchers had been claiming all along: that these scientific ideas were, in fact, now distant from the average mind, issuing from on high, both magnificent and fundamental and therefore requiring translation into a vernacular format that excited and amused its listeners as much as it informed them.

In the long run, one of Agassiz's true achievements was to help firmly root the Academist image and rhetoric in American science, to attach the highest cultural value to the ideal of basic research. His status as "that grand foreigner" allowed him to accomplish what none of the other Lazzaroni could, that is, to make an elitist science appealing on seemingly populist grounds. If his own science proved to be a failure, based on a static view of species as finalized "thoughts of the Creator," if his rather outmoded drive to collect and classify was itself matched by an equal need for accumulating public attention and large amounts of money for his own private use, these factors were less important in the long run than Agassiz's coup in putting research on an equal footing with invention for the middle and upper classes of the mid-19th century.

SCIENCE AND HIGHER EDUCATION: CHANGE WITHOUT CHANGE

The late 1820s saw the first period of major curriculum reform in the older colleges. This reform, centered on the acceptance of more science (and, to a lesser degree, modern languages and literature), was a direct response to Practicalist demand and a related decline in enrollments. The soul-searching among college trustees and presidents at the time was not especially profound and by no means radical (recall the 1828 Yale Report, discussed in Chapter 2). The essential Academist belief that a liberal education must be kept "free from all study of a practical kind" never wavered an inch. Adding a course in geology here, a semester of chemistry or natural history there, meant little or nothing in terms of compromising established Academist priorities. Thus, perhaps inevitably, as an answer to social realities and attitudes, this type of change proved to be an utter failure. It was wholly insufficient both with regard to the public's new love affair with technology, which gained force from the 1830s onward, and with regard to the call for increasing support for research and full professional status among the scientific professoriat itself.

Indeed, at a fundamental level, the "reforms" of the 1820s were a kind of facade to hide a lack of deeper change. Two decades later a majority of the older schools remained no less focused on humanistic studies as the core of their program, and thus, given the rapid changes that had occurred in a new industrializing America, they were even more disconnected and out of step with certain strains of the contemporary sensibility. This situation, in fact, was diagnosed with consistent vehemence and intent by none other than one of the more well-known college presidents of the time, Francis Wayland, who had come to Brown in 1827 and there, for the next two decades, applied his best efforts toward expanding the technical curriculum, yet who met with no small resistance along the way from trustees and professors alike. Early in the 1840s, gazing over a national scenery replete with "steam, machinery, and commerce," Wayland wrote that "the College or University forms no integral and necessary part of the social system. . . . It plods on its weary way solitary and in darkness," uninterested in the "active professions" to which its potential students were now turning. "In no other country is the whole plan for the instruction of the young so entirely disservered from connexions with any thing else" (1842, p. 41).

This is not to say that the Academist program of the old colleges was entirely out of touch with the times. As we saw in the previous chapter, a fair portion of America's intellectual and political elite—including social reformers such as Horace Mann—continued to hold fast to similar beliefs even while employing science as a reservoir of utilitarian rhetorical imagery. Such men, who had themselves been trained in the Academist faith, were obviously unwilling to abandon the humanities as the model and standard for any true education. This would have amounted to self-denial, a turning away from shared values regarding "wisdom" and "moral purpose." Education was, most of all, a matter of high ideals. Science fit these ideals only on an adjunct basis, as something to add on at the edges, never so as to displace or otherwise mitigate the holy centrality of the classics.

Bolstered by loyalty to such ideals among prominent nona-cademics, the colleges no doubt felt all the more justified in the ethic of their stance. In some part, their stand against expanding the curriculum was perceived in terms of saving "true culture" from the demands of "King Mob." Wisdom could not allow itself to be compromised or diluted. Moreover, the country needed to retain the only institution capable of producing its higher in-dividuals, leaders infused with the noblest thoughts, lessons, and discipline culled from the greatest epochs of the past.

Yet two realities could not be ignored and eventually im-posed change in spite of everything else. The first involved a new round of public criticism, which once again affected enroll-ments. The colleges responded by trying to lower tuition and board, keep down the cost of books, and even offer temporary financial waivers. This, however, had little affect, except to bring many schools even nearer to the brink of financial disaster. "We have produced an article for which the demand is diminish-ing" wrote President Wayland. "We sell it at less than cost. . . . We give it away, and still the demand diminishes," (1842, p. 44). A second influential phenomenon was the advent of active patronage for scientific training by private donors, who were themselves caught up in the new spirit and were being urged by academic scientists to make their belief tangible by providing important sums of money to found chairs, programs, and even colleges in the sciences. This was a potential source of bounty that the older schools, in their economic fragility, could not pass up. In addition to survival considerations, it should be stressed that arguments for change, specifically for more science, came from within academia itself, from students, professors, and pre-sidents other than Wayland, who all perceived the existing curriculum, fixed as it was on Greek, Latin, rhetoric, logic, and moral philosophy, as woefully disconnected from the new em-phasis on progress. The fact that ambitious students of science were obliged to seek training in the universities of Europe, particularly to learn experimental procedure and methods of research, was cause for shame on all fronts. All of this too combined to form a key logic for change, a nearly uninterrupted call for the colleges to play catch-up through the 1840s and

1850s. This they did, most certainly. By 1850 nearly every school had broadened its offerings to include courses in geology, mineralogy, physics, chemistry, and astronomy, and some had gone still further by establishing professorships of biology, zoology, and botany. According to some estimates, by this time a typical faculty of nine or ten members commonly included three professors of science and at least one of mathematics (see, e.g., Guralnick, 1975a, p. 116). At Brown and at Williams, under the reform-minded presidencies of Wayland and Mark Hopkins, the ratio was even higher, as much as seven-to-three. Courses taught in chemistry and mineralogy, meanwhile, were touted as fulfilling the popular demand for training in practical subjects, and the consulting activities of professors in these fields as well as those in geology were held up as examples of how the college now directly served national commerce. Also by this time, the old recitation system was itself undergoing reform: Partly in response to the European example, partly to the new prestige of research, and partly to professors' understanding that they were engaged, to some degree, in training apprentices, many science classes began to employ a mixture of lecture and experimental demonstrations. Actual student participation became much more common.

Thus it might seem that a true transformation took place in American academia during the final two decades of the antebellum period. But this would be a misinterpretation. A closer look reveals many limitations to the character and depth of this episode of change. One of the most striking of these, from today's perspective, is the universal refusal to adopt a system of electives, such that students could more fully pursue separate intellectual careers. Instead, all new science courses were either simply added to the required curriculum, or divided off into a so-called Modern Department that also included history, modern languages, English literature, and philosophy (this program was the true forerunner of what is today known as the liberal arts, yet then was considered by most trustees and classicists as a necessary compromise, lower in status to the time-honored, largely medieval, Greco-Latin curriculum). Either way, reform provided an alternative without really offering choice. At Brown, President Wayland claimed that the curriculum had become so "crowded" that a free elective system was absolutely necessary if the college was to continue to grow; yet his pleas went unanswered in his own college and were opposed at others. Taking stock of the situation during 1855, the influential *North American Review* (Vol. 80, p. 151) argued not for an approach such as Wayland's, nor for some other flexible alternative, but instead for historical regression, a complete return to those "permanent studies which have been regarded by wise men, for a thousand years, as the best possible discipline for young minds."

The lack of an elective system meant that a limit existed to the amount of science a college could conceivably incorporate. It also helped ensure that scientific courses would be added more or less peripherally, as a series of branches grafted onto an established trunk. At Brown, despite four professors on a seven-man faculty teaching science and mathematics, the hours devoted to these subjects numbered only 160 out of a total 479 course hours for the undergraduate degree (a mere third). Here, as elsewhere, teaching was still entirely book-based, focused on the reading and learning of one textbook for an entire discipline. In a single year (usually the junior year), the course of natural science would include the study of elementary texts in as many as seven different fields, from astronomy and chemistry to geology and zoology.

This single course would be the only exposure the student would have to any of these disciplines, each of which was roughly comparable, in terms of time and effort spent, to a single Latin or Greek text. Finally, despite changes in teaching methods, there was still no provision made at any college for students to perform actual laboratory research, to become trained in what was already viewed as true professional science. Well into the 1850s the established intellectual Silk Route to Europe went on uninterrupted (even at the verge of the Civil War, Germany and Scotland had trained more than half of America's most prominent researchers, this being countered only at Rennselaer by Amos Eaton, whose "learn-by-doing" philosophy was ridiculed by the trustees of most colleges even as it was envied by other scientists). And while teaching methods may have altered somewhat, examinations—decisive points in any student's career—did not, being still a matter of verbal performance based almost purely on rote recall from lecture and textbook material. Clearly, despite important institutional reforms in some areas, the larger substance of educational practice continued to be based on the same fundamental premises during the entire antebellum period.

Indeed, those positive changes that did occur were largely countered by others that acted to protect the reign of Academism. The most significant innovation here, beginning at Yale in the late 1840s, involved the creation of separate scientific schools, where technical training of a professional kind could be given, yet kept wholly isolated from the college proper. Housed in separate buildings, allowed to set up programs that included both practical and laboratory training, these scientific schools and their students were segregated from the traditional colleges and their student bodies. They existed on an entirely adjunct basis, without financial support from the college proper and, still more significantly, without the right to grant degrees. At Yale, the initial proposal for work in applied science came mainly from Benjamin Silliman, one of the college's most prominent professors. Silliman sought endorsement by the Yale Corporation for the appointment of two new chairs, one in agricultural chemistry, another in "vegetable and animal physiology." Given that Silliman had located a benefactor who promised $5000 if $20,000 more could be raised, the trustees agreed, issuing a resolution that made sure the burden of fundraising rested outside their purview and that the two-year, nondegree program was limited to offering instruction "to graduates and others not members of the undergraduate class" (in Fulton & Thomson, 1947, p. 210). The two professors to be hired were to pay the college a fee of $150 per year for the use of college facilities (a building then vacant) and had to supply all their own equipment. On Silliman's recommendation, the corporation expanded the original science program to allow for the teaching of other, nonscientific subjects such as philosophy, literature, and history, but only so long as these did not in any way overlap the given, traditional curriculum in theology, medicine, and law. Launched in 1847, the program was more substantially underwritten during the mid-1850s by Joseph E. Sheffield (father-in-law to one of its professors), and was dubbed the Sheffield Scientific School in 1861.

In effect, the Sheffield School offered the nearest thing to the first graduate program in America—a program centered on science education, involving students directly in research work. Yet it was a program that began without benefit of true academic standing, in a disenfranchised school, whose faculty was denied the right to confer a recognized degree on any of its students. This was to be amended; by the beginning of the Civil War, the success of the Sheffield Scientific School, and the growing prestige of the scientific community generally, forced the college to endow Sheffield with full privileges. In 1863 it conferred one of the nation's first Ph.D.'s on J. Willard Gibbs, perhaps the most prominent and important American chemist of the 19th century. But this later alteration does not diminish the fact that the Yale Corporation originally viewed the program as a way to isolate and restrict any further expansion of science at the college. Founding an auxillary, self-supporting scientific school was an acceptable means of answering the call to modernize while leaving Old World priorities intact. Indeed, so clear was the message at the time, that on dedicating the program Silliman felt it was his diplomatic duty to state openly that "the students having access to [the school's] advantages will be strictly professional and not academical" (in Miller, 1970, p. 91). These would be students, in other words, of the Sheffield School and not of Yale College.

This policy of noninclusion was employed elsewhere. Almost simultaneously, Benjamin Pierce at Harvard secured funding for the Lawrence Scientific School, intended, like Sheffield, as a professional (read: vocational) program "for the use of young men who desire a scientific education without the diploma of the college and without the classics." Several years later, in 1852, Dartmouth established its Chandler Scientific School (with a starting grant of $50,000 from Abiel Chandler). Separate nondegree technical schools and programs were thereafter begun at a host of other colleges, including the universities of Virginia, Michigan, and North Carolina.

These schools were a first step; at least they brought a real version of contemporary science within the sacred gates. But they did so only by forcing scientists, in a sense, to continue to accept an auxillary status and position. Science could finally be an academic profession; it had achieved its own brand of Academism, being granted both a dimension of higher truth and of Practicalist Franklinian training. But to those who still held power in the world of higher learning, science education continued to be accorded second-rate treatment.

BENJAMIN SILLIMAN: STATESMAN OF SCIENCE IN THE NEW REPUBLIC

While debate still surrounds the Lazzaroni, regarding their influence, there is unanimous agreement among historians regarding the pivotal significance of Benjamin Silliman (1779–1864), whose efforts to establish academic science in America during the first half of the 19th century had no rival. Some hints of his importance, and of Silliman's career, have been given above. While his actual contributions to scientific knowledge were less than profound (consisting of a dozen or so articles in geology and mineralogy), his impact as an educator, an administrator, and an editor—and therefore as a type of well-rounded ambassador from the Scientific Community to the rest of academia—probably did more to secure an institutional future for American science than all the important research efforts among his contemporaries. More than any other individual, Silliman became, from the late 1820s onward, the trainer, promoter, and the exemplary academic model for the new type of scientific professional who would take command in America by the 1840s and 1850s.

Silliman himself was the most unlikely of reformers. He came from a Puritan New England family of affluent means. His father and grandfather were Yale-trained lawyers, "gentlemen of the old school," his mother the daughter of a prominent minister, educated at Harvard. Nicknamed "sober Ben" while still in his teens, Silliman was raised on the Bible and the classics, and remained devoutly religious his entire life. As a young man, he had strong Federalist sympathies, was an ardent anti-Jeffersonian, and a strong believer in the distinctly patrician way of life passed down to him through his forebears, whom he was groomed to succeed. Graduating from Yale in 1796, he went on to study law until 1801, when he was suddenly, unexpectedly, offered a chair in chemistry by his friend, Timothy Dwight, then president of Yale.

Half out of deference, Silliman accepted the post. It was a move that fixed the course of his life, that in some sense severed immediate obediences to his past yet allowed him to rediscover them in another form. In any case, out of necessity he embarked on a pilgrimage to Philadelphia, to New Jersey, and eventually to Europe, eager to acquire the education he would need to fulfill his duties as Yale's first professor of natural science. Time spent at the University of Edinburgh deeply impressed him as to the scale, popularity, and support scientific training received on the other side of the Atlantic. It swayed him away from chemistry and toward geology and mineralogy, the "study of God's earth," courses that, in addition to chemistry, he eventually established at Yale after his return. In his early years at the college he helped found the medical school and updated its curriculum to include courses in chemical and pharmacology. Between 1810 and 1812 he procured for Yale the large and exemplary mineral collection of Colonel George Gibbs, this at a time when such collections were nearly unknown yet crucial as a base for studying earth materials. Silliman also made sure that the Gibbs Cabinet, consisting of nearly 10,000 specimens, was open to the public, both as a kind of tourist attraction and as an educational opportunity. He wrote, "It kindled the enthusiasm of the students and excited the admiration of intelligent strangers, [helping make] New Haven . . . the focus of travel between the North and the South" (in Fulton & Thomson, 1947, p. 86).

Silliman's founding of the *American Journal of Science and Arts* (1818) was noted in the previous chapter. Within less than a decade, it became the most prestigious scientific periodical in the country—the first, and for some time the only, truly national journal where technical research found a consistent, privileging forum, yet where it also coexisted with philosophy and the arts under a larger educational intent (thus, in a sense, providing an early model in print for the later scientific schools, and their

version of the "liberal arts"). Some idea of the journal's importance, in fact, can be gleaned from Silliman's own reflections
at the end of his first 10 years as editor:

> The editor has been frequently solicited, both in public and private,
> to make [his journal] more miscellaneous, that it might be more
> acceptable to the intelligent and well educated man who does not
> cultivate science; but he has never lost sight of his great object,
> which was to produce and concentrate original American effort in
> science, and thus he has foregone pecuniary returns. . . . Others
> would not have him admit anything that is not strictly and tech-
> nically scientific; and would make this journal for mere professors and
> amateurs. . . . But our savants, unless they would be, not only the
> exclusive admirers but the sole purchasers of their own works, must
> permit a little of the graceful drapery of general literature to flow
> around the cold statues of science. (in Fulton & Thomson, 1947, pp.
> 126–127)

Lyrical conciliation toward literature notwithstanding (it
was, after all, the *American Journal of Science and Arts*), Silliman
seems already to have become an important, recognized broker
of status on the periodical scene, widely identified with the
image of a national science, able to genteelly reprimand those
who had already taken on the attitude that scientific work was a
rarefied enterprise with little or no interest in public understanding.

Silliman's opposition to this attitude, at least in its pure
form, grew directly out of his own theological view of science, a
view widely shared by educated Americans of the time and
especially typical of scientists engaged in the fields of natural
history. Simply put, Silliman believed that science constituted a
way for human beings to gain knowledge of God through study of
his works, and by means of such knowledge to receive lessons in
moral restraint, in reverence, and appreciation, and also, on the
practical side, new God-granted powers to satisfy the physical
needs of humanity. Science involved doing God's work on earth
and therefore should be made available, in some form (hopefully
educational), to a great many Americans. For Silliman himself,
at some level, his own teaching was quite literally a form of
missionary work, a matter of bringing students closer to the

divine. While he never expressed it this way exactly, the degree of devotion he granted to his pedagogic efforts, the care and concern he gave his students, the magisterial, even pulpit quality of his lectures, and the variety of his activities regarding the dissemination of scientific knowledge—all blended neatly and expressively with the Puritan fervor to which he remained loyal all his life.

By the 1850's, after a full half-century of work, Silliman himself became a subject of mixed views among scientists. The Lazzaroni, for example saw the journal as too "diluted" with nonscientific, nonresearch matter, and privately spoke of its editor's own failure to produce important discoveries in his chosen field. During this last decade of his life, he was viewed by the new researchers as something of a relic: At the 1856 meeting of the AAAS, on the event of dedicating a new geological museum and reviewing the history of geology in America, Silliman, known to many as the "venerable Bede" of academic geology, was not even asked to speak.[2] Three years later, at another annual meeting, Bache offered to compensate this impropriety by proposing the following resolution: "That this Association express its gratification at the presence of our octogenarian friend at our meetings and we hope that we may often meet him, as he now is, with his eye undimmed and his force unabated" (Fulton & Thomson, 1947, p. 257). The venerable Bede, in short, was already consigned to the mists of history. Yet the irony in such condescension was apparent to many even at the time. Bache and Co. depended almost entirely upon, and were devoted to furthering, the prestige of academic science, which no one had done more to solidify than Silliman. No doubt it was important for them, as a younger generation interested in claiming the future as their own, to reject the past as being "unformed." Yet their Academist rhetoric, their hopes for a truly American science on a par with that of Europe, their emphasis on professionalism, were all largely derivative, made possible by Silliman's more quiet yet patulous influence.

This influence, in fact, was probably greatest in terms of the students converted to science, whose careers the Yale professor did much to establish. An early example was Josiah Holbrook, who went on to found the Lyceum movement. Others included many of the most successful researchers and science educators in antebellum America: Amos Eaton at Rensselaer, Edward Hitchcock at Amherst College, George T. Bowen at Nashville University, James Dwight Dana at Yale, Chester Dewey at Williams, and Denison Olmsted at the College of North Carolina, to name but a few. It was Silliman too who helped persuade Daniel Coit Gilman (then librarian at Yale) and Charles W. Eliot of the importance of science to the founding of a true university, these two men who were later to assume transformative power in American higher education as the reform-minded presidents of Johns Hopkins and Harvard. Beginning in the mid-1830s, meanwhile, as a result of his renown as an academic lecturer, Silliman himself was encouraged to join the lyceum circuit, at a time when prominent scientists were still a rare phenomenon as public speakers in America. Having seen such speakers in Edinburgh, and desirous of producing an effect similar to the worldy admixture of entertainment and dense pedagogy that had so impressed him, Silliman adapted his own classroom methods to the purpose, preparing his lectures with great care, weaving in numerous chemical experiments (rehearsed to ensure their smooth working), expansive geologic drawings (e.g., five-foot-square panels showing scenes of the primeval past), gleaming mineral specimens, and impressive fossils—all devices which set the acknowledged standard for public lectures in science thereafter in the United States. Indeed, it could be said that Agassiz too, with his great panoramic drawings as a backdrop and his broad array of paleontologic samples, was merely following the lead established by Silliman a generation earlier.

Beyond these things, historians have also pointed to the fact that the Yale professor "virtually created the field of scientific consulting, mainly in the mining industry, and was instrumental in transforming this activity into a recognizably professional one, which then took on an institutional form in the Sheffield Scientific School" (Brown, 1989, pp. xii–xiv). The significance of Sheffield as an early graduate school and as the offspring of Silliman's own efforts has been noted. Endorsement of the school would no doubt have been much less forthcoming were it not for Silliman's own national reputation, his long-term proven ability to attract students and fame to Yale College, and the quiet, but very real, power he wielded among its trustees. Indeed, he had been the presiding officer of the board of examiners (whose function was to evaluate student applicants) for decades. Looking back on the 1830s and 1840s, Theodore Woolsey, then president of Yale, gave a fair account of Silliman's value as a measured asset:

> His personal presence, his great popularity, his fine powers of persuasion caused him to be put forward whenever there were wants to be urged before the legislature or before private friends, whenever strangers of distinction were to be honored, whenever . . . responses were due from the authorities of the Institution. . . . He, in his prime, was our standing orator, the principal medium between those who dwelt in the academic shade and the great public. (in Fulton & Thomson, 1947, p. 162)

The conclusion to be drawn from all this is a crucial one for the early history of American science education. Silliman made one of his most critical contributions by helping ensure, through his classroom practices, his editorial work, his public speaking, and his diplomacy, that professional science would have its dominant residence in the university and therefore would be attached to teaching. He was a man who might be said to have appeared at just the right historical moment, when science required an ambassador of his qualities.

But what were these qualities, at the last? Again, Woolsey provides a clue: "His influence was all exerted in favor of discipline and order . . . he was a tower of strength to the government" (in Fulton & Thompson, 1947, pp. 161–162). A recent biographer, Chandler Brown, meanwhile, makes the point differently, claiming for Silliman an abiding desire, above all else, to "conserve a way of life" defined by the values of old world character, dignity, elevated purpose, and Puritan social ideals (1989, p. xv). At the time Silliman joined the Yale faculty, this way of life was still ambient among those with means in New England. A generation later, when Jacksonian resistance and accusation had risen from "below," he was already deeply ensconced in what remained the most conservative academic institution of higher learning in the nation. Even by that time, the new president, Jeremiah Day, decided proponent of the old Academism, was a friend and associate and saw the now middle-aged representative of science at Yale as a compatriot in mind.

Silliman, in short, was less a reformer than an adapter and modifier. His beliefs about the meaning of science and its place in the world placed him in a position of philosophical compromise. He could easily accept both his elevated position as an "orator" for the college, yet the unequal and comparatively minor status of science within it. No doubt he did, in fact, have "his face turned toward a future in which science would stand side by side with the classics and theology" (Fulton & Thomson, 1947, p. 273). But the emphasis here should be placed on "future," for even at the dedication of the Sheffield School in 1847, he was still perfectly willing to agree that scientific work remain outside the medieval nexus of the college curriculum, therefore in its old kneeling position before crumbling altars. Indeed, the point would seem to be that Silliman himself, as the figurehead of Yale science in whom so much was apparently invested over the years, provided the college with a rationale for remaining largely content with what it had, for not expanding its scientific program too widely, too quickly, too far beyond the vision of their man Silliman. This is not at all to call him a lackey or puppet; such would be absurd. Yet, except perhaps for Sheffield, the medical school, and the Gibbs mineral collection, Silliman's greatest work for science was done beyond the borders of Yale, where his students became ardent reformers who worked to expand technical training in America. It is instructive, in itself, that in an age of so much turmoil and development, when it became the norm for so many of these students, as capable scientists, to change positions several times in the course of their careers, seeking posts where their freedom for scientific work might be the greatest, Benjamin Silliman, intellectual patron of them all, stayed right where he was, in the town and college where he had been living since his teens, and where his forefathers had lived for generations past.

What seems clear is that only a man of Silliman's sensibilities could have succeeded to such a degree in making a larger place for science in early antebellum academia, bound to tradition as it was. Only a professor with the old Academist bearing and learning, who proposed himself as no reformer and no great innovator, could effect the acceptance, if even on a supplementary basis, of expanded technical training associated with the old colleges. It is, perhaps, a characteristic of the Jacksonian era that such conservatism as Silliman's, in following its own logic across the decades, seeking to do something of God's work in a new way within an older institution, ended up producing a series of radical innovations. Silliman was no Emersonian leader, no researcher of power or eminence. He probably felt as much at home with the theologians who comprised the Yale Corporation as he did with the scientists who attended the first meeting of the AAAS. In any case, by the time of the latter, he was himself a veritable icon or monument of renown. In that same year (1846), for example, he appears in Nathaniel Hawthorne's "A Virtuoso's Collection," a story intended as a spoof of the learned mind, cast in the form of a natural history museum. Hawthorne, in fact, narrates a humorous visit through this collection comprising relics from all the classical works of literature, including the taxidermal specimens of Minerva's owl, Puss in Boots, and Noah's dove, along with literary weapons such as Don Quixote's lance and the bow of Ulysses ("unstrung for ages"), and items ranging from the salt that had once been Lot's wife to the cup from which Socrates drank the fatal hemlock. Near the middle of this seemingly endless catalogue of the contents of the well-taught mind, this sentence occurs: "There was a small vase of oracular gas from Delphos, which, I trust, will be submitted to the scientific analysis of Professor Silliman" (Hawthorne, 1982, p. 705). For the man often dubbed the "father of American science education," this placement is perhaps appropriate, a kind of proper literary burial in an age fast coming to a close.

Science as "Culture":
Education and Modernism
in the Late 19th Century

Science and education, as topics of national consciousness, have both tended to emerge in America during periods of deep uncertainty. The topography of their importance has been to provide elevated visions of order and consensus, and also of a more progressive, egalitarian society. But they have also been seen to give reasons for dreams of renewal and reform and for the very conditions of their possibility. In the late 19th century, especially after 1870, these conditions were sorely needed. And they were met, on many fronts, by a reverse excess that planted "science" on the face of nearly every claim to truth. The United States, that is, endured another crisis of self-confidence; the uncertainties, indeed, were considerable. As noted by Cremin (1988, p. 1),

> Americans commemorated the centennial of the Republic in 1876 with a curious mixture of pride, pretension, cynicism, and shame. There was no escaping the fact that the country was in the throes of a depression, that its government was ridden by scandal, that its southern states continued under a military occupation, that its army was at war with the Indians, that its newly emanicipated blacks were neither enfranchised nor literate, and that some of its most articulate women had chosen the Fourth of July itself to promulgate a Declarations of Rights for Women. . . . Yet, there was the boost to the

national spirit that derived from the great Philadelphia exposition, which was attended by one in five Americans and which confidently proclaimed American industrial preeminence to the world.

Industrialism in the United States, in fact, continued to develop almost unabated throughout the whole second half of the century. The Civil War aided this tremendously—not so much by jolting the North into a period of technological expansion (this was already well underway), but by largely destroying the agrarian economy of the South. Economic collapse in the South forced many people to migrate into the cities, to find work in factory, mine, and other industrial settings. By the late 1880s, fully a third of the nation's people (roughly 21 million) were living in urban areas, which entered their second great era of expansion a half-century after Jackson. As before, this urban growth was stimulated by immigration, which brought millions of mainly poor Eastern and Southern Europeans to the swelling metropolitan centers.

The new cities were an environment of confusion and struggle. Enormous differences in wealth and comfort came to be omnipresent. New forms of faceless, collective human existence—the so-termed *mobile vulgus*—became the norm. In nonurban areas, church and family remained important institutions, but the culture of rural life, formerly so central to the American experience, was passing away. The railroads, the expanding road and highway system, the advent of modern communications via the telegraph and mass-circulation newspapers and magazines, and, of course, the spread of compulsory schooling, had opened up and ended the isolation of small-town America. A good deal of this had been underway since the 1840s (see, e.g., Larkin, 1988, pp. 204–231), but it intensified enormously following the Civil War. The great medium of "the world," intruding everywhere, weakened local frames of mind and replaced them more and more with a public, national identity of Americanness, replete with urban and industrial anonymity. Directly expressive of this, late 19th-century Americans were drawn to a deep interest in an idealized vision of their colonial past, as a time of ritualized harmony and a slower, simpler, more secure pace of life centered on hearth and home. Other, more intense reactions to the new urbanized society took place when the "gilded age" collapsed into severe economic depression and mass unemployment near the end of the century, in the years 1893 to 1895. Widespread agrarian discontent precipitated by this depression, and the resulting groundswell 1896 presidential campaign of William Jennings Bryan, with its ardent anti-intellectualism (directed in some measure against science, biology in particular), was in large measure an attempt to reassert religious pastoralism. But the desire to reclaim a lost world of small-town life and rural virtue, whether expressed through a love of pre-Revolutionary antiques or through an effort to install fundamentalist sympathies in the White House, was doomed. In its failed bid to turn back the tide of "demon progress," it revealed more than challenged the triumph of metropolitan existence, and with it, the "faceless crowd" (as Edgar Allen Poe once called it) in the new industrial America.

Within this general climate, then, science was associated with new uses and new realities. Between the Centennial Celebration of 1876 and the Bryan campaign of 1896 a new kind of science emerged, a science correlatively stripped of most of its former pastoral connections, irreversibly associated with types of knowledge, application, and a real-world power now at the heart of an urban "Western Civilization." The change was profound, and it was to have equally profound effects. For with the passing of bucolic science—a science of smaller scale and accessibility— the image of the individual researcher and inventor also faded away, to be replaced by "big science." In the 19th century "big science" meant university science, whose work carried out an undertaking of major institutional organization, supported by such entities as major corporations and the federal government, who thereby shared in guiding scientific training and setting scientific agendas.

The search for answers to both intellectual and social questions became a critical part of the new image of science. Before the end of the century science had become a kind of "sacred canopy" (to adopt a phrase from cultural anthropologist Peter Berger, 1967) under which to place, explain, and dissolve nearly every uncertainty of the age. Given the powers and power invested in it, "the scientific" could not but help acquire messianic capabilities.

HERBERT SPENCER: SCIENCE, SOCIETY, AND THE ROLE OF EDUCATION

Louis Agassiz for a time strode the popular scene as a scientific colossus and cosmopolitan, gathering funds and earning prestige for almost any project he decided to promote. Despite what he managed to accomplish for the "new researchers" and the image of science generally, however, his glory was short-lived. It was soon seen that his science was outmoded, a matter of pre-Darwinian conceptions in a gaining Darwinian age. His effect upon the curriculum, both at the college and secondary school levels, was important in securing a place for biology; outside this, however, his impact was minimal. By the 1870s he was gone, and the torch of scientific renown had been placed in the hand of another, one whose impact would be far more profound and lasting.

Herbert Spencer (1820–1903) was the great popularizer of Darwinian theory in the United States. This does not mean that he championed all Darwin's ideas. Rather, he took what he wanted from the Darwinian corpus and applied it toward creating a "science of society" and, more specifically, a new kind of education. Unlike Thomas Huxley, who recognized the limits of evolutionary theory, Spencer believed evolution provided the answer to human existence. His career of prodigious work and vast influence was based on relentless pursuit of a few analogies, adoped with near abandon.

For Spencer, society itself was an organism. Change and progress came about through "the never-ceasing action of circumstances upon men—by the constant pressure of their new circumstances upon them," he wrote in Social Statics (1850, p. 384). The purpose of education, like that of history, was to promote "adaptation, self-preservation," which, for the individual, meant preparation for "complete living"—for health, profitable work, civic duty, moral behavior, and enjoyment, all the things that would best advance "the forces which are slowly civilizing us" and which the traditional curriculum had not, and could not, adequately teach (p. 384). The question of "What Knowledge Is of Most Worth?" had an absolute, unequivocal answer:

> The uniform reply is—Science. This is the verdict on all the counts. For direct self preservation, or the maintenance of life and health, the all-important knowledge is—Science. For that indirect self preservation which we call gaining a livelihood, the knowledge of greatest value is—Science. For the due discharge of parental functions, the proper guidance is to be found only in—Science. For that interpretation of national life, past and present . . . the indispensable key is—Science. Alike for the most perfect production and highest enjoyment of art in all its forms, the needful preparation is still— Science. And for the purposes of discipline—intellectual, moral, religious—the most efficient study is, once more—Science. (Spencer, 1860, pp. 84–85)

The salvation of human beings lay in the power of scientific knowledge to deliver certainties about every aspect of thought and behavior. No society that expected to survive the century could ignore these "truths." No nation that claimed progress and freedom as its aim could linger in putting them to work. Spencer not only inverted the "liberal arts" in terms of power relations, he made the humanities a kind of subset of the sciences (thus, in effect, predicting their demise as the givers of "wisdom" for "good living").

Regarding the learning process itself, he had an equally encompassing theory of "natural education." This he derived not from Darwin but from an older notion, given recent form by Haeckel's famous epigram—"Ontogeny is the brief and rapid recapitulation of phylogeny"—which, in Spencerian terms, became the process by which individual mental development rehearsed that of the human species over the course of its history.[1] By means of this analogy, Spencer gave the concept a new importance, suggesting utilitarian possibilities that implied the curriculum itself could be designed "according to nature." If learning was a process of natural laws, then the education of children could be structured to follow those laws, perhaps even to enhance them. Scientific in activity, learning could also be

scientific in all its contents. Spencer did not advocate a mechanical or fixed teaching method. On the contrary, he said, "Children should be led to make their own investigations, and to draw
their own inferences. . . . They should be *told* as little as possible,
and induced to *discover* as much as possible" (1860, pp. 124–
125). This method would best allow the natural process of
"individual evolution" to take place. Yet, in social terms, Spencer's language implied something else: The true model for learning in the classroom was not nature but the scientiest himself,
whose intellectual work, characterized by patient search and
revelation, represented the highest form of evolutionary advance
within society.

Spencer's ideas about science were Academist. Science for
him, as for all positivistic thinkers, was the realm of absolute
truth and heroic truth seeking. In this, he helped greatly to
spread the gospel of the new researchers, beyond even the
influence of Agassiz. But he did this more through his explanations of society and his theories of education—perhaps his best
known writings in America—than his discussions of nature itself. Indeed, Spencer's larger appeal—and his notions of education as well—cannot be separated from his theory of social
Darwinism, which seemed to explain and justify the existing
social order of middle-class prosperity, upper-class privilege, outright racism, and widespread industrial exploitation of the
masses. Here, Spencer's "science" became pure biodeterminism:
He argued that economic divisions were "natural" divisions, an
elemental expression of the "struggle" for fitness. The mech-
anisms were fixed and certain; social intervention was wrong, for
it amounted to tinkering with the inevitable. Education itself
thus had to be seen, in the final view, as a way to help partition
the classes, to make them adapt to their true station in life and
immediate circumstance. The learning of science would aid such
understanding. It would teach the necessity of the social order
and would thus help the lower classes "prepare for life."

Spencer's works were widely read in the late 1860s and had
an enormous influence in shaping the intellectual climate of the
1870s. What they evoked, in a mixture of Victorian eloquence
and biblical pronouncement, was a vision of science as the
critical requirement for comprehending and building civiliza-

tion. The nation that did not train its young scientifically would fail, just as surely as a machine built according to faulty principles or an organism deprived of nourishment. Promoted by the tireless E. L. Youmans (himself a frequent writer and lecturer on "modern science"), by such contrasting apostles as William Graham Sumner (social Darwinist *par excellence*) and Lester Frank Ward (who believed society could be redirected and improved through education), Spencer entered the fray regarding a classical education versus a science-based education like a storm. Within less than a generation, his ideas prompted three very different movements. The recapituation hypothesis inspired G. Stanley Hall to promote child study as a basis for developing a truly scientific curriculum. The notion of "efficient study," meanwhile, became a calling card for industrial involvement in education and helped encourage the use of F. W. Taylor's "scientific management" as a model for educational practice. Finally, Spencer's belief in classroom "discovery" anticipated the pedagogy of John Dewey, whose ideas for a child-centered process of learning became so central to Progressive education as a whole.

Between 1870 and 1890, meanwhile, the combined influence of grand Europeans such as Agassiz and Spencer, with their appeals to American nationalist vanity and belief in progress, helped make science an object of popular interest. Spurred by the fame of both, a new, far more passionate amateur involvement in natural history developed among the middle and upper classes. This interest was further stimulated by continued exploration of the West and reports of this exploration carried in newspapers, magazines, and journals—for example, of Clarence King's geological forays along the 40th parallel; John Wesley Powell's harrowing trip through the Grand Canyon; and, perhaps (at the time) most famous of all, the "great dinosaur wars" fought between Edward D. Cope and Othneil C. Marsh during the years 1877 to 1888. The West now became more than ever a near-mythic region, saturated with all manner of national destinies, located somewhere between the savage and the civilized, but always a great planting ground of "godly science" in raw form.

Renewed Enthusiasm for Technology

The nationalistic side to social Darwinism had an interesting counterpart. This took the form of a curiously inverted pastoralism, transformed from the technoswoon of the 1840s and 1850s. The late-19th-century industrial boom in the United States was founded, above all, on natural resources: oil, coal, iron, gold, silver, timber. These were the literal substance of the "national body," whose vast quantities and the wealth to be made from exploiting them were essential to the fulfillment of "manifest destiny." Newspapers of the late 19th century were filled with organic metaphors that were transferred from technology to its "parent" creator, industry. The new factories were characterized as "born on invention," "fed by labor," in need of "constant growth," "nourishing" to the rest of society.

Public enthusiasm for technology, in fact, had waned only with the coming of the Civil War. Beginning with a new episode of rapid invention, and with Spencerism as a whole, the 1880s and 1890s saw the fervor pick up again in force and continue well into the first decades of the next century. The rapidity of industrial expansion would not have been possible without a concomitant wave of innovations in engineering. New, more liberal patent laws, allowing for greater inventor ownership, helped encourage a burst of activity in this area. Of the 600,000 or so patents granted during the whole of the 19th century, over 535,000 were granted after 1865 (Rippa, 1992, p. 130). As before, the idea of "invention" became central to the sense of "American genius." Modernism and progress were imagined in the concrete form of new devices, this time on a far more massive and international scale than before, including such things as the telephone, the "wireless," the cable car, the bicycle, the repeating rifle, the cinema, the automobile, electric lighting, to name but a few. The announcement of each new discovery promised huge, sometimes overwhelming benefits to American living, and was commonly offered in language reminiscent of the technological sublime. Yet something else had been added, for now inventions were often described in technical detail, as if to a discerning audience of active amateurs.[2]

This gave the impression, false to be sure, that technology was a kind of public property, to be shared and celebrated by all. As before, what was in fact only a shared discourse of declared benefit became a veil to hide the harsh, often appalling realities that made such technology possible. "Science," as the promise of abundance, was employed to disguise science as a practice aimed at profit, national power, and imperialism.

These associations between technical knowledge and American progress surfaced early on at two major celebratory events, both held in Philadelphia: the 1874 World's Fair and the 1876 Centennial Exposition, which together have often been labeled the inaugural ceremonies of the Gilded Age. The first of these drew significant attention from educators for its display of European industrial schools and their "hands-on" curriculum. The Centennial, meanwhile, of far greater scale, was a direct attempt to link technical power and national progress as essential elements of the American self-image. Its focus was not history or the arts, but instead "a century of American progress" defined in technological terms. Its scale was overwhelming: a site of 240 acres, 167 buildings, as many as 30,000 exhibits from some 35 nations, centered on a main exhibition building deliberately erected to be "the largest human structure in the world" and dedicated solely to "the achievements of national science and industry." It was here that new technologies such as Bell's telephone, Edison's telegraph, the Singer sewing machine, and the typewriter went on display as examples of the "American national genius." They were seen by more than 8 million visitors. Science—once again, as a sign of success for all of American society—had the power to draw the nation together, to both intellectually explain and materially advance the given fabric of life and thereby overcome even the divisions wrought by war. Such was the Centennial's deeper message. As Kasson (1976, p. 233) has noted, it was a message that, in the decades to come, "continued to pay tribute to republican ideals while supporting a social system that effectively denied their promise . . . a system that converted republicanism from an animating ideology to a static buttress of the conservative industrial order."

CHILD STUDY AND EDUCATIONAL EFFICIENCY

In *The Age of the Crowd*, Serge Moscovici (1985, p. 13) writes that "The birth of a form of collective life has always coincided with the appearance of a new type of human being." In late 19th-century America, this new human being was the child. The child, that is, as the embodiment of mass hope for individual successes of many kinds; the child, too, as a vessel for good intentions, so often based on ideas of what science could show to be good and needed for growth; finally, the child as a personification of science itself, in the form of a "new psychology" that would be capable of documenting and proving what it meant to be "human."

In the United States, periods of rupture and uncertainty tend to promote a reexamination of the future in both personal and national terms. In the second half of the 19th century, such reexamination focused a great deal upon children, both as the real and symbolic embodiments of this future. Childhood, in fact, was in the process of being redefined, mostly in romantic terms, as the happiest, purest, and most innocent stage of human life. As a corollary to this, children also became the vessels into which all of society's doubts and fears might be projected. The troubled status of the traditional family was by now a matter of public record: The increase in rates of divorce, crime, and urban poverty; the breakup of neighborhoods and dissipation of former communities; the struggle for women's rights—all these had become overt issues in the press and helped bring the question of the child, as a carrier of frailty and hope, to the fore of public consciousness. Ideas of the child's place in nature and the general drift of social disruption came together. As Karin Calvert notes (1992, pp. 138–139):

> Many scientists [psychologists, doctors, sociologists] of the nineteenth century were less than sanguine about America's future. They found ample evidence of children inheriting the vices and deficiencies of previous generations, but much less indication that positive qualities were as frequently transmitted. Study after study concluded that great men rarely produced great sons, but that weak and corrupt men habitually sired tainted offspring. . . . Where

Americans in the years after the Revolution had looked forward to the ultimate perfection of mankind and had seen their own children as part of the process, mid- [and late] nineteenth-century parents feared the cumulative corruption that surrounded them and strove to protect their children from its influence.

The child was the "precious inheritance" of nature, but also (and therefore), potentially at least, the revelation of a degenerate world. However, hope did exist for children to overcome this reality, including even a poor hereditary beginning. That hope lay in the idea of "environment," that is the choice and construction of settings capable of "suppress[ing] the weaknesses and cultivat[ing] the virtues inherent in [the] child's nature" (Calvert, 1992, p. 139). Part of the child's innocence was thus his or her malleability: If shaped and aimed in the right direction early on, by "proper training," he or she would almost certainly escape the evils, the "contaminations," that otherwise lay in wait.

Given this sensibility, which became fully ambient for the middle class after about 1870, one could expect that childrearing guides would also become hugely popular. In contrast to earlier periods, many of these guides were written by doctors who claimed knowledge of "the science of childhood" rather than merely the benefits of experience and worldly wisdom. Parents were urged to follow methods derived from "the fruits of science," which tended to ordain love and affection in some circumstances (e.g., to encourage interest in learning) but strict discipline and forbiddance in others (e.g., masturbation). Manuals governing diet, exercise, hygiene, vocabulary, etiquette, and so on, were frequent best-sellers.

But such aids to domestic education were not the most fundamental change that took place in accordance with the new view of childhood. This came about instead with regard to institutional education, where the child's relationship to nature—or rather, the nature now thought to exist in the child—was used as a basis for a logic that might discover the most natural, progressive, and successful type of program for learning. Several groups of academic writers, for example, applied evolutionary ideas to the child, emphasizing aspects of growth and imitation. Their first cause came back to the recapitulation hypothesis, which they used, with Spencerian abandon, to model the forging of a "scientific curriculum." Two groups in particular adopted this idea: G. Stanely Hall and his followers; and the Herbartians (see Kliebard, 1992).

For these two camps, the path seemed clear and open. Spencer had naturalized society and industry, and education, too, as an institution. Now was the time to fully naturalize the child and the curriculum. Put differently, as the vessel of nature, the child became the "father" of the individual and therefore the agent of change for society. As a way to evoke and train this nature, the school and its curriculum became the "environment" for remediation, the agent of change for the child. According to this logic, the curriculum came both first and last. The goal appeared to be a simple one, to use study of the child's natural psychological development as the basis for designing a course of training, one that might best advance such development. Yet the reality of translating "science" into design proved complex, muddled, all the more so when this "science" was itself problematic.

Hall's notion of research, in fact, was a vestige from an earlier age. He sought to build enormous, specimen-type collections of data on "the contents of children's minds," for example, how they cleaned themselves, what they liked to eat, what phrases they used, and so on. The idea was trenchantly Baconian: If enough facts were gathered, generalizations would emerge. That they did not was of no surprise to the many psychologists interested in experimental work (including John Dewey), who eventually came to see Hall as a zealous pretender to the throne of a "scientific psychology." Despite this, however, as president of Clark University, Hall was a vigorous popularizer of his own ideas and a powerful force in pedagogy before 1910 (he was also responsible for bringing Freud to America in 1909 and giving him an honorary degree).

The Herbartians, on the other hand, were a group of important educators, including Charles De Garmo, Nicholas Butler, Frank McMurry, Charles McMurry, and Joseph Rice, who came together in the early 1890s for purposes similar to those of G. Stanley Hall. Inspired by the early psychologist and romantic philosopher Johann Friedrich Herbart (1776–1841), these men

developed the recapitulation hypothesis into a notion of evolving "cultural epochs." Briefly put, their idea was that the "principle of succession" in human history involved a staged advance
through distinct periods of cultural—not biological—activity
(e.g., the "hunting epoch"). The history of each such epoch
should be taught to children of a specific age, so that progress
from year to year would condense, in exact and literal parallel,
the movement from past to present of the "race" as a whole. Like
Hall's theory, this view had a certain mystical character to it, a
whiff of teleology. Even by the late 1890s, it too was coming
under criticism, again by those such as Dewey (who had been a
member of the group earlier on). Part of the problem, ironically
enough, was that despite its scientific pretensions, the theory
strongly favored a classicist view of "culture," with humanities
and the arts preeminent. It argued for "correlation" between
subjects on the basis of a unifying theme; yet literature tended to
be the "core" around which all these subjects of learning would
be integrated into a whole.

The larger impact of Herbartism, in concert with Hall's, was
extremely significant—indeed, it was more than significant: It
was foundational. Taking their cue from Spencer, these two
camps established a new mythology in America with regard to
the link between education and science. From this time on, it
became standard practice to use scientific information or
methods in curriculum design. Indeed, the very idea of "designing" a curriculum became a fixed assumption. Whether the hope
was to enhance a student's natural development, his or her life
goals, the needs of society, or projected "competencies," education was now to be inseparable from science (such hopes, in fact,
were themselves dependent on this merger). The scientization of
the child had helped lead to a scientization of learning; and this,
in turn, would soon help foster a scientization of the school, as a
bureaucratic institution.

On the face of things, it might seem highly ironic that little
of this injection of science into curriculum making overturned or
disturbed the basic Academist order of subjects. Major changes
were made, to be sure; scientific subjects were granted a larger
place. In secondary schools across the country, for example,
physics, biology, chemistry, and geography, were increasingly

incorporated into the standard program after 1870. Yet, as before, these subjects remained very much second to the humanities overall (English and foreign languages in particular). Most
educators continued to favor the image of their own intellectual
past. Despite their hopes for change, they were also very much
part of a conservative perception that saw a "lowbrow" mass
inhabiting the cities, a raw material to be shaped and elevated
for the good of society. The urban student was everywhere seen
by these men in terms of a grand conversion, a potential
transformation into the "highbrow" individual—cultured, liter-
ate, self-reliant, and capable. The democratic faith invested in
education lay exactly in this presumed capability, earlier evident
in those such as Horace Mann, to disseminate the elitism of
former times—and to do it "scientifically." And yet the idea of
"science" was also being used here to define a new cadre of
professional experts. Only the new "educational engineers"
could create the magnificent Academist individual in mass form.

The new power granted to science, a power of pedagogical
truth and method, was thus not to be a reason for educating
children themselves in science. As the vessel of benefit, the
child did not need to understand the subjects and principles
upon which his or her training depended. Moreover, whatever
"correlations" or "discoveries" were to occur within the classroom, whatever "interests" were to be set in motion, the process
of learning was to take place according to fixed laws and rules
defined strictly by adults, not in ways that would allow children
to be the creators of their own learning, the true and active
centers of education. It is in this way, then, that the curriculum
can be seen to have taken precedence, as a molding force.

Indeed, the recapitulation analogy is striking when looked
at a bit more closely. It rests on the belief that children are
primitives who must be gradually "civilized," stage by stage, so
that they may one day reach the level of "high culture." To those
who do poorly or who leave school at an early age there clings
the suggestion that they are less civilized than others, almost
"natural" representatives of a more backward historical order.

Child study was used to mount a direct attack upon the
prevailing mode of recitational teaching in the schools. Yet,
while it helped to reconceptualize instruction, its impact on

classroom method was relatively small. Part of the reason for this was that it shared a fundamental view that underlay this older system. It argued that nature should be replaced by culture, in an orderly, controlled fashion. Or, in the words of William Torrey Harris, the most prominent conservative educator of the time, "Education is the process of the adoption of the social order in place of one's mere animal caprice" (Harris, 1898, p. 2). This was perhaps the crudest of all traditional Academist rationale, focussed as it was on the simultaneous training and shaping of the individual self in a generic mold. It was also the oldest of such rationale, reaching back for its inspiration to Plato's *Republic*. It remained effective in late 19th-century America partly because of how it was redefined in Spencerian terms—that is, "education" as the redirecting or productive channeling of natural impulses—but also because Hall and the Herbartians gave it a new and more "sensitive" look.[3]

HIGHER EDUCATION: THE RISE OF THE RESEARCH UNIVERSITY

> You will please keep in mind that this is a college and not a technical school. The students who come here are not to be trained as chemists or geologists or physicists. They are to be taught the great fundamental truths of all sciences. The object aimed at is culture, not practical knowledge. (in Noble, 1977, pp. 24–25)

This kind of statement, delivered at midcentury by school officials of Williams college to a newly hired chemistry professor, Ira Remsen, remained possible for decades thereafter, and the spirit within it, still longer. The great changes that had taken place in public images of science, regarding both its Academist and Practicalist nature, and the forced beginnings of acceptance within higher learning generally, did not lead in any immediate way to correlative adoption in the older colleges. That adoption, on a full scale, would finally come about as the result of two phenomena: first, the rise to fame of Charles W. Eliot, who introduced radical reforms at Harvard after 1870; second, and perhaps more important, the rapid development of the research university in the 1880s and 1890s—a development so remarkable for its scale and swiftness that it would come to mark, for some time, the decline of the older colleges as the centers of American higher education.

And yet even then change came slowly to these schools, as the old order collapsed but refused to die. Harvard, in fact, helped set the pace for resistance as much for reform. When appointed president in 1869, Eliot proclaimed in his inaugural address that henceforth "This university recognizes no real antagonism between literature and science, and consents to no such narrow alternatives as mathematics or classics, science or metaphysics. We would have them all, and at their best" (in Hofstadter & Smith, 1961, vol. 2, p. 602). But the gap between this modernizing intent and actual attitudes among academics remained strong. Indeed, "antagonism," with its assumed equality of status, misrepresented the truth of the situation. For during the 1870s, 1880s, and early 1890s, Academism as a philosophy of humanities "culture" actually became more influential in the older colleges. This resurgence of Academism was sustained, in general, by the conservative mood of the time among the affluent, and by several things in specific: (1) the increased status of "Ivy League"-type schools among the wealthy, whose own personal values were oriented less toward scholarship than social finish, and European "worldliness"; (2) the related founding of expensive prep schools and private academies such as Exeter, Choate, Groton, and so on, typically carried out by classicist expatriates who were often reacting against the "low standards" of the new public high schools, and who therefore wanted to create precollege colleges based on the Academist ideal—these exclusive private schools acted as guarantors of entry into the older colleges for the sons of the *haut bourgeoisie;* and (3) the founding of many new technical schools after 1870, as a result of the Morrill Act, which essentially protected the older colleges from having to make too many important reforms in the direction of science and Practicalist demands.

For these and other reasons, the old colleges became in large part Academist playgrounds for the patrician class. Indeed, as pointed out by Cremin (1988, p. 158), when, as a young man, William T. Harris entered Yale in the late 1850s, intent on studying the so-called "three moderns"—modern literature, modern science, and modern history—he discovered none of them yet in existence there. Within the borders of the older schools, science made few institutional advances of a truly profound kind until the late 1880s and 1890s. Instead it was still cloistered mainly within the adjunct scientific schools. Indeed, almost any "modernization" was confined to these same schools. It is noteworthy, for instance, that the Sheffield School at Yale openly competed for students against Yale College (the terms "school" and "college" indicate the differences in presumed status)—not the least because by the 1860s the former had developed a prototype curriculum for today's undergraduate study, involving not only the natural sciences, but modern foreign languages and literature (including English), and the new social sciences as well. Increasingly, those with ambitions similar to Harris' were drawn to Sheffield and repelled by the college. As Nathan Reingold points out (1991, see chap. 9), it was exactly this split between an updated course of study and the recalcitrant classicism of the college that prevented Yale from becoming a true university until well after 1900.

At Harvard too, though Eliot went to work immediately and with great vigor to dismantle "old fogeyism" (as it was called), to make Harvard the modern model for American higher education, his reforms took over 25 years to complete, by which time it was simply too late. In that span of time, the scepter of academic power had been seized by the new research universities, which rapidly supplanted the old colleges by becoming the nuclei of knowledge production, most of all scientific knowledge production. Eliot himself was chosen in part for his declared ambitions to make war on the past: He was young, only 35; he was controversial; he demanded a purely "American college," one not based on European models; and he himself was a scientist, a chemist and mathematician. His most fundamental reform, in fact, might well be seen as an act of revenge. It was, after all, a program of attrition, of reducing one by one, year by year, battle by battle, all the old sacred subjects into electives.[4] During this time the curriculum was broadened greatly and tough entrance exams were implemented. The Lowell Scientific School was effectively absorbed into the college. Professional and graduate schools were developed and put on a par with one another.

And yet, such changes should not be misinterpreted. Eliot was a reformer but no Reformist. Indeed, he merely shifted the terms of Academism in turn with the new era. He may have angered many of the old guard during his long tenure, but his fundamental view of higher learning was conservative. He thought of education as "a natural outgrowth . . . and an expression of the average aims and ambitions of the better-educated classes" (in Cremin, 1988, p. 381). A liberal patrician, Eliot believed keenly, even anxiously, in "the cultivated man," someone whose knowledge would now expand beyond classicism to include "the prodigious variety of fruits of the imagination that the last century has given to our race" (Cremin 1988, p. 385). Regarding what such cultivation would provide, meanwhile, Eliot included not only standard democratic ideals, but a tightly unified national community and a near-total faith in expertise as well. He noted, "Confidence in experts and a willingness to employ them and abide by their decisions, are among the best signs of intelligence . . . and in any democracy which is to thrive, this respect and confidence must be felt strongly by a majority of the people" (Cremin, 1988, p. 384). Democracy, in short, would "thrive" only on the basis of intellectual plutocracy. And who better to supply the new plutocrat of the mind than the new college, Eliot's own brain child?

Eliot's reforms, therefore, were actually a kind of compromise. By no means did they signify an overturn of former pieties. Though he rejected classicism in its details, he adopted its spirit in modified form. That he became, in the last years of his life, creator and editor of the famous "Five-Foot Shelf" of Harvard Classics is hardly a surprise.[5] Here, "science"—in the form of its own rhetorical monarchs—could itself be brought within the standard Academist notion of "great ideas."

Elitism among the classicists had helped keep the professoriat, by choice and by default, removed from the realm of public influence. The schools were at this time controlled by their boards of trustees, comprised mainly of members of the clergy, who were the ardent spokesmen for tradition. During the

1870s these men began to be replaced by the new power brokers of urban society, lawyers, bankers, and businessmen, those who could give or raise money for the college. This change formed one part of what Hofstadter (1963) has called "the status revolution," whose larger ambit involved the rise of academic scholars to the level of public spokesmen, reformers, and "experts." Some part of this new willingness to "go public" probably derived from a new confidence in themselves regarding their own power as knowledge-holders in a knowledge-based world. Historians have also pointed to the sense of "humiliation" the professoriat often felt with regard to the new "captains of industry," who held so much real-world power yet who so often ridiculed classical learning for its "bookish" irrelevance. By the mid-1890s more and more scholars were taking the new "titans" to task for their "savage" disregard of democratic values, human needs, and social betterment. But even beyond these things was the rise of "science" as the final intellectual capital for useful expertise.

In 1862, in the middle of the Civil War, Congress passed the Morrill Act, perhaps the single most influential move ever taken by the federal government with regard to higher education. It was a move, at one and the same time, that led to the founding of the research university and to the direct—but selective—injection of state control over portions of higher learning.[6] It was, in particular, a decisive and massive underwriting of science, "pure" and "applied," that rapidly changed the entire face of American education and helped propel the country within a single generation into the forefront of modern intellectual culture. The act had been first proposed in 1859 by Justin Morrill, a self-taught, self-made Congressman from Vermont, whose lifelong goal it was to help democratize higher education along Practicalist lines: in particular, to "aid the industrial classes in the procurement of an education that might exalt their usefulness" (in Nevins, 1962, p. 4). His idea was to provide a subsidy to every state in the form of a grant of land, amounting to 30,000 acres for each representative and senator a state sent to Congress, this land to be devoted to the founding of a college. "The leading object" of such a school, meanwhile, "without excluding other scientific and classical studies, and including military tactics, [would be] to teach such

branches of learning as are related to agriculture and the mechanical arts . . . in order to promote the liberal and practical
education of the industrial classes in the several pursuits and
professions in life" (in Knight & Hall, 1951, p. 542). Clearly, it
was the classics that fared least within this scheme.[7]

When the bill was first introduced in 1859, it was vetoed by
President James Buchanan on several grounds. Among these
were included the fear that it interfered directly with state's
rights, that is, it constituted a move to make the states, via
education, overly dependent on the federal government. Buchanan and other conservatives also held it "extremely doubtful . . .
whether [the] bill would contribute to the advancement of agriculture and the mechanical arts," and this because it would
"injuriously interfere with existing colleges . . . in many of which
agriculture is taught as a science and in all of which it ought to
be so taught" (in Knight & Hall, 1951, p. 285). In a gross
misreading of his own time, Buchanan stated, "No father would
incur the expense of sending a son to one of these institutions for
the sole purpose of making him a scientific farmer or mechanic"
(pp. 285–286). This blatantly misstated the truth of things,
ignoring as it did the rising success of the new scientific schools
such as Sheffield, the expanding number of technical colleges
like Cooper Union, the general celebration of technology then
ongoing. It assumed that the only worthy education—worthy,
that is, for a father to consider giving his son—was that of the
older colleges, still largely following their classicist bent. New
institutions of higher learning, professional in outlook, would
"injuriously interfere" both on economic and philosophical
grounds.

The Civil War, however, and a new president had changed
much. Lincoln signed the bill, understanding at once its
possibilities to aid in rebuilding the Union by supplying each
state with trained, professional personnel. Under the act, states
were allowed to sell whatever land they did not intend to use for
the actual building of a college. The proceeds of such sale,
meanwhile, were mandated to be "applied to the uses and purposes prescribed in [the] act," that is, on the college.

This infusion of land and capital quickly led to the founding
of state colleges (soon to be universities) throughout the coun-

try. At first, many of these institutions, in their initial phases, stuck fast to the "A and M" letter of the act. During the 1870s and early 1880s, they remained largely vocational in emphasis, focussed on the training of exactly the kind of scientific farmers and mechanics that Buchanan had thought so antinomial to a "higher" education. Soon after, however, these new schools began to shift their programs in a broader direction, one meant to encompass more of this traditional "higher learning."

This shift came about due to the impressive and formidible examples set by Daniel Coit Gilman of Johns Hopkins and Andrew Dickson White at Cornell, two college presidents who decided early on to build their institutions in accord with the most advanced demands for progress of the age—demands that highlighted research and "the pursuit of new knowledge." On the whole, these men consciously adapted certain features of German (Prussian) universities to their cause, in particular those that favored the increase of knowledge over its transmission. Even by the middle of the century these universities had become the envy of the western world, both for their devotion to basic research and, no less important, the magnaminous amount of government support they received for this purpose.[8] In some sense, it was Gilman and White's mission to Americanize the German model, open it up, make it a more potentially democratic institution, yet all the while retaining its elevated self-image, as a place where the epistemological future began.

For the sciences, therefore, the Morrill Act and the example of Gilman and White proved a boon. White turned Cornell into the undergraduate university par excellence. He instituted the "breadth and depth" curriculum in which the sciences were the solid equal of the humanities, and in which science students were required to perform their own experiments. He set a new standard by opening the doors fully to women, and by giving both alumni and faculty a place on the all-powerful board of trustees. Though he lacked the funds to establish complete laboratories in every field, he still provided for faculty research and scholarship in many areas and ensured that none of the technical disciplines went begging. Gilman, meanwhile, focussed his efforts at Johns Hopkins on graduate study. He did not require federal monies to establish his program: With an endowment of $3.5 million in 1876 (remarkable for the time), he fashioned within a mere decade the prototype graduate institution, devoted to research and publication almost exclusively. As a diplomat-educator, eager to set trends, he was especially canny in his use of an older, establishment rhetoric to support an entirely new kind of school. "The object of the university is to develop character—to make men," he said; "It should prepare for the service of society a class of students who will be wise, thoughtful, and progressive guides" (Gilman, 1898, p. 33). Yet his real object, as revealed by his practice, was to create a new coterie of research workers, utterly free of accountability to outside (Practicalist) demands. And he did this by eliminating the undergraduate program altogether ("the traditional 4-year class system" he called it) and in its place creating a "factory for discovery." Gilman understood that research had no meaning if it was not published. Thus, Johns Hopkins led the way in founding a host of new academic journals covering a wide range of technical fields. In the late 1880s it began publishing all Ph.D. dissertations through its own university press, and thus by 1900 had transformed itself into an enviable intellectual force, radioactive with "discovery."

The new land-grant schools tried to combine the examples of Cornell and Johns Hopkins. The laboratory, as a site of "pure knowledge" production, was elevated in terms of its prestige in the academic community. By this time, the fame of those such as Louis Pasteur and Robert Koch had already transformed the life sciences and much of natural history into new areas of experimentation and quantification. Furthermore, again on the German model, the end of the 1880s saw a growing number of corporations in the chemical, mining, and steel industries establishing their own in-house research labs and technical training schools. Science writers, meanwhile, had identified the laboratory for the public as the "forge" of technical work. Inventors such as Edison began to abandon the older term "workshop," tied as it was to suggestions of craft in a pretechnologic age, and began to speak of "laboratories for discovering the future" and of the "invention factory" (a phrase that speaks of Edison's own ambitions to make himself the economic equal of the new business titans). Whether for basic or applied work, the lab was seen as a "citadel of creation," whose physical removal from the workaday world became the guarantor of a purifying distance from society and a proximity to truth (Hughes, 1989). Finally, it was no accident therefore that the laboratory soon became a useful metaphor for any number of determinisms, as in discussions of human society as "nature's laboratory"; a presidential campaign as "our laboratory of politics"; war itself as "history's laboratory." One is reminded of how the idea of the "American experiment" had been similarly employed several generations earlier. Yet the lab image implied a different sort of institutional stamp: It suggested not merely removal, but removal as the condition for wrestling with and working out some immanent, fundamental principle of nature, a struggle whose results and meanings were therefore beyond the reach (and responsibility) of human control. It was an image, then, that sometimes paid respect to social Darwinian notions.

The cause of the new university gained further help from another source too. The risen status of these institutions, and their accepted claim to supplying American democracy with its new leaders, helped encourage patronage of a whole different order by the new industrial elite. By the middle 1890s, when criticism was growing against the monstrous effects of robber barony, many plutocrats warmed to the idea of founding a university as a means to display good will toward the public and the nation, or else (in the case of Andrew Carnegie) to offer atonement for a lifetime of unbridled greed and acquisition. Thus, John D. Rockefeller donated a staggering $30 million for the founding of the University of Chicago; Cornelius Vanderbildt offered $1 million to found a school in his name; and Andrew Carnegie set up his vast system of research institutes, libraries, and foundations. Most of this money was intended, specifically or by association, for science: for basic research, applied engineering, invention, and industrial training. Like the federal support effort it was intended to both mimic and outdo, this new corporate philanthropy signified more than just the concrete politicization of science: It proved that technical knowledge was now itself a "natural resource," something to be fought for and paid for.

SCIENCE AT THE CENTURY'S END

The ascent of research promoted by the Lazzaroni, Spencer, and others, had come to pass. By the end of the century science had been elevated to new levels of authority, both within the academy and in society at large. Indeed, as a new intellectual model, it went well beyond the Spencerian limit. Wedded to destiny through evolution, "science" spawned a host of new offspring. Darwinian ideas, quantitative methods, the concepts of "fact" and "objectivity" began to be applied to nearly every discipline in the humanities. This invasion of the old "liberal arts" by ideas of "the scientific" was in no small way responsible for the birth, growth, and spreading influence of the social sciences in America: economics as a science, scientific sociology, experimental psychology, political science, and scientific history.

Such developments within the academy reflected public aspiration. At this time science seemed to offer a return to certain Enlightenment ideals, in revised form. In place of the speculative unity of all knowledge, science offered unity derived from method (involving induction, evidence, and theory). In place of the slow climb to human perfectibility, there was now hope in the possibility of multiple liberations—from false knowledge; from superstition; from poverty; from inequality; from disease; from inefficiency; from poor education; from national weakness. Science had come to fulfill, in brief, the role of prophecy. To groups otherwise antinomial in outlook, to the public as a whole, it offered abundant hope, a promise of deliverance through the revelation of fact and the teachings of truth. In an age of disturbance, it became the sign of the new as a promise of order, certainty, and faith.

None of this, however, meant a change in the basic rule of Academist sensibilities. The acceptance of science, in idea and practice, did not at all end or weaken this reign with regard to ideas of education, especially at the college level. On the contrary, it made for the very opposite. For now, higher learning was higher than ever, more in the business of creating elite leaders, increasingly dubbed "experts," than even in the past. And while this had occurred through decline of an older idea of "the cultured individual," such an idea had itself been allowed to continue through its very required opposition to the new and more potent Academist ideal—the concept of science as *Wissenschaft*, the search for those "fundamental principles" that superseded all utilitarian demands (yet that might have their fertile applications, all the same). It was an ideal taken up alike by otherwise conservative historians such as Henry Adams or by reformers such as John Dewey. The scientific ideal was the newfound grail by which intellectuals of all sorts could claim, at one and the same time, the removal of "disinterest" and the proximity of the "expert."

In reality, therefore, the "liberal arts" philosophy did not become any more flexible during this period. Rather, its power merely shifted, under new management. American scientists, after all, in particular those like the Lazzaroni, had from the beginning sought a diplomat's revolution in the curriculum, and had succeeded in part because of it. They had argued for acceptance by adopting the language of conventional academic power, the existing language of elitist removal. Over time, this strategy gained a grounding in public attitudes and federal financial support. For a variety of reasons, people were won over to the purist view—indeed, their beliefs and desires with regard to progress depended inexorably upon it. Following those such as Bache and Henry, a sequential hierarchy had been firmly built in the minds of nearly everyone: First came "discovery," the wresting from nature of basic knowledge; then came "invention," with its genius of selective application; finally, there was a "product," available to the many, whether in the form of electricity or the zipper, which then transformed daily life.

This was the pattern that people believed now propelled the present into the future, a pattern that was taking place more rapidly and successfully in America than anywhere else. It was the identical pattern, in its trickle-down Practicalism, to the idea of "method" and "nature" in the social sciences, whose discovery of natural laws could be used to develop new practices, for example, of curriculum design or teaching approach, that in turn would transform the given local context of human activity.

Whether the product was a machine, a procedure, or a student, the basic process was the same.[9]

Academism triumphed in this belief. By the 1880s and 1890s, "science" in academe could be easily separated from "technology," as the origin of new knowledge, the first step in the process. Scientists were not engineers; their inquiry and experimentation were no longer seen as having any *necessary* ends beyond truth. But this was enough; truth once generated could be used by others. Engineers, in the meantime, were being made into scientists. Their knowledge might have had practical aims, but it nonetheless had to begin with more fundamental laws and principles. Thus, any split between the Practicalist, utilitarian image of science and the academic view existed only on the literal level. Both shared the same idea of "pure" science. The conflation of perspective was potent: In the end, science would serve America and produce inventions because it *was* objective, removed, elitist. Only the most fundamental truths could build the most powerful technologies. Only the most basic science could be the best for America, its economy, its schools, its eminence. Thus did the manager, educator, and professor speak different dialects of a single heroic language.

The main arguments for curriculum reform in American higher education, therefore, continued to focus on science through the end of the century. Such reform moved forward on the back of many demands, pushed ahead by a range of competing and overlapping rationales from many sources, public and private. And in this the movement was wholly successful. Indeed, both complex and formative, the effort as a whole had a single aim: that more science be included—more science, in more variable and specialized and prestigious forms. By the first decade of the 20th century, America had truly joined the "first nations of science." And science itself, as method and knowledge, had become not merely an important element in education, but the very model for its progressive reformulation.

Science and the Progressives: Standards and Standard Bearers in the Age of Reformism

Between roughly 1830 and 1890 the United States went from being a republic of farms and mills to a nation of factories and offices. In some form, therefore, it could not but change correspondingly from a landscape of local schooling, private and domestic, to vast educational "systems" at the secondary and university levels. Throughout the country, children and adolescents became the new intellectual laboring class; teachers, therefore, gained in measure of respect, scrutiny, and criticism. The rise of education was predicated on the rise of science—as idea, as source of imagery, as a profession, and as a reconstituting ideal. This double rise was itself dependent upon the ascent of the middle classes and therefore on middle-class anxieties and demands in the late 19th century.

Two events in 1893 had sweeping effects in terms of changing public attitudes. The first of these was the Panic of 1893, which led in the following years to economic depression, mass unemployment, and worker unrest. This event impacted on many aspects of American life and caused a widespread failure of belief in the idea of inevitable progress that had been so tirelessly promoted by laissez-faire industry and government. The resulting social chaos seemed to lay bare, finally and for all to see, the terrible inequalities of American society, the betrayal by corporate avarice and political corruption of American virtues, such as civic duty, moral responsibility, and democratic individualism.

More and more people began to believe that society could not be left to itself; it must be taken in hand. Thus, by the late 1890s, when a new era of prosperity had begun (powered in large part by technological advances), the "business before people" mentality was actively denounced. A call for new communities, a new revitalization of "democratic living" was proclaimed from every quarter. In such a climate of desired change and recognized means, it was almost certain that attention would turn to the schools, to learning.

The second event, adding focus to the cry for reform, was the publication of a single book—Joseph Rice's *The Public School System of the United States*—which created a national sensation. Rice was a pediatrician who had spent time studying pedagogy in Germany under the famous psychologist Wilhelm Wundt, and who had come back to the United States ardently convinced of the need for a detailed "science of education." At the behest of the *Forum*, a well-known educational journal, he conducted a survey of classroom practices and schoolboard administrations in some thirty-six major cities. What he discovered, in the words of Cremin (1988, p. 227), was a national phenomenon of "rote learning, mindless teaching, administrative ineptitude, political chicanery, and public apathy." Despite exceptions of excellence in a few areas, the overwhelming evidence revealed "waste" and "drudgery" on every side. It was time, said Rice, for a new beginning, for the founding of a true "progressive school," based on the ideas of science. Rice's book set off a period of intense soul-searching among the public as well as the pedagogic community. It helped make the school the symbol of nearly everything that had gone wrong with America—and therefore the place to begin reform. As such, Rice's work became the oracle of its own influence, the rallying point from which the Progressive movement in education began.

THE EFFICIENCY DEMAND

Though Progressivism in general meant demands for change, these demands were tremendously varied and often in conflict with one another. By no means were they always radical of even liberal in spirit. Indeed, the movement as a whole was deeply tied to the rise of positivism in America, and as such, contained many conservative elements.[1] For every writer such as John Dewey, there was another of more modest insurgency like G. Stanley Hall (whose influence persisted), and still others, such as Edward Thorndike and Louis Terman, whose sensibilities were conservative. For every reformer who envisioned a new kind of education centered on the child, there was another for whom the curriculum was the major concern, and yet others which condensed upon the "scientific" as a warrant for total standardization. The first two of these positions frequently called upon "nature" and argued that reform in learning would bring a better democracy. The last camp, meanwhile, talked of "scientific management," and while thoroughly devoted to a better America, defined "better" not in terms of individual fulfillment but rather in terms of functional adjustment and organization. Overlap existed between these points of view, either in general rationale or in the detailed recommendations offered. And no small reason for this was the general status granted to "science," which, in one form or another, ran through nearly every program of the time.

Scientific knowledge and reasoning, in fact, had by this time been everywhere taken up as the obligations of truth and objectivity. Technology seemed proof of this as a legitimate belief. It implied that intellectual problems could be understood in material terms through science, above all in the image of the machine, as a perfect organizational system. Science, in brief, could sooner or later be applied to any area in which knowledge and its uses were important. In the early 1880s Edison wrote of electric lighting that "Like any other machine, the failure of one part to cooperate properly with the other part disorganizes the whole and renders it inoperative for the purpose intended" (in Hughes, 1981, p. 124). Two decades later one finds much the same words applied to social institutions, the national economy, human personality, and the curriculum.

By the 1890s people were calling for, or actually proposing, ideas more fool-proof, productive, even automatic methods of work—whether this work be steelmaking or teaching. In education, the Herbartians argued that the traditional curriculum had been "wasteful" because "unnatural." This stance shared affinities with a petition for increased "efficiency" in schooling propounded by a growing cadre of utilitarian-minded industrialists, politicians, and social theorists on the right, all of whom took technology as their model for reform. These interests formed a distinctly unified and influential voice that had continuing effects as late as the 1930s. It was a powerful voice, despite public criticism of its sources, because it appealed time and again to Practicalist sympathies for change, pleading for an education more immediately responsive to the needs of the work force and the national economy. Students could be trained "efficiently" as future workers at many different levels of expertise, and this could be done in a factory-like manner, with a minimum of waste to produce a maximum of "product." Schools could save the nation millions of dollars annually by taking on the responsibility of preparing children for their future life's labor, something that would advance progress and national power to new and unknown heights.

As a whole, this movement had two basic effects. First, it helped promote the rise of vocational education as a new element in the secondary school curriculum. This was important, for education urged by reformers on the left, who asked that learning be more in touch with social needs and therefore that it make a break with traditional Academist elitism. A second effect, however, was yet more significant: The call for "efficiency" gave new vitality and direction to the concept of scientific curriculum making by aiming it not at intellectual virtue but at functional utility—the goal being to teach the best and most useful knowledge in the least amount of time with the least amount of effort. While the child and the curriculum were being made more scientific, the watchword "waste" was therefore being used to demand a parallel change in the school and classroom. In the decade after 1910, this would come to involve adopting as an organizational model F. W. Taylor's "principles of scientific management," written expressedly to prove "the great loss which the whole country is suffering through inefficiency in almost all of our daily acts," and to show how this might be remedied by a penetrative use of "systematic management," itself "a true science, resting upon clearly defined laws, rules, and principles."

This was an argument that harked back to the technological sublime. It envisioned a school built not merely as a system but as a literal machine. It involved a structure of rationalized bureaucracy ("working parts"), which would perfectly achieve the mechanized organization of human activity. In the words of Ellwood P. Cubberley, one of the most influential educators of the time:

> Every manufacturing establishment that turns out a standard product . . . maintains a force of efficiency experts to study methods of procedure and to measure and test the output of its works. . . . Our schools are, in a sense, factories in which the raw products (children) are to be shaped and fashioned into products to meet the various demands of life. . . . This demands good tools, specialized machinery, continuous measurement of production . . . the elimination of waste in manufacture, and a large variety in the output. (1916, p. 338)

The factory/machine analogy had the power to cast everything in a cold clear light. Every element of schooling had its place, its function, its immediate goal, and its measurable result, all of which could be designed and operated appropriately, by an equally utilitarian "expert." What lay at the base of everything was a sense not merely of waste but of failure—the failure of the individual. For those such as Horace Mann, the building of institutions as "systems" had been thought necessary to supplant the lost efficacy of the family as transmitter of necessary virtue. For the new efficiency experts, however, this system was not enough by itself: It left too much to chance, to the unpredictable choices and personalities of the people themselves, as people. What was required was a total system, saturated with accountability (testing most of all), in which both teachers and students could be "mass produced" and therefore guaranteed in terms of their "product quality."

Thus, seemingly from another direction altogether than that of child study, "science" again became a kind of runaway analogy, used to transform nearly every aspect of schooling into a form of technology. Yet, efficiency study was perhaps not so divorced from child study as has sometimes been assumed. The guarantee of "the scientific," after all, whether for Cubberly or G. Stanley Hall, was felt to lie in the possibility of erecting a program by which a better individual (more natural, more efficient) could be produced *en masse*, through a design that incarnated inviolable laws, rules, and principles. It was this aspect of science, the Academist idea of final truth wrested from nature, a truth then applicable to Practicalist ends, that had appeared so attractice. And the child? Again, to neither Hall, the Herbartians, nor the new efficiency experts, was this entity the true origin or active force in learning. As both the container and the receptor of laws and principles, the child was no less the "raw material" and "finished product" for all of them. The final irony, therefore, lay in this: that in seeking to protect the child from a world of dehumanizing force and scale, child study ended up giving aid to the discourse of its enemies, by projecting a central part of the rationalized world they rejected inside their very subject, thus making it, as intellectual object, a leading implement in its own loss of possible power. The movement for "scientific management" in the schools, with its concomitant need for standardized exams, standardized selection and evaluation procedures for teachers, and standardized methods of curriculum building, succeeded because of its attractions for many different interests both without and within the educational world as it already stood in the early 1900s, a world hungry for "the scientific," a world that child study had helped build.

John Dewey and the Liberal Turn

As pointed out by H. M. Kliebard (1992), John Dewey had offered the most farsighted and effective critique of such thinking about the "scientific school." In 1901 Dewey wrote, "It is easy to fall into the habit of regarding the mechanics of school organization and administration as something comparatively external . . . to educational purposes and ideals." He then went on:

> We think of the grouping of children in classes, the arrangement of grades, the machinery by which the course of study is made out and laid down . . . as, in a way, matters of mere practical convenience and expediency. We forget that it is precisely such things as these that really control . . . the personal and face-to-face contact of teacher and child. The conditions that underlie and regulate this contact dominate the educational situation. (pp. 337–338)

Dewey thus saw that the school had already become a kind of rationalized technosystem, with a wholeness to it that reverberated at every level of responsibility and action, not the least in how it "regulated" human contact. To try and alter one part of it—say, the classroom or the curriculum—while leaving the rest intact would doom any reform to failure, no matter how well intentioned. In fact, the more radical the reform, the more inevitable and profound the failure it would meet, since the more disconnected it would be from the actual workings of school "machinery."

The point is an extremely powerful one, applicable to any era and any kind of education. The lyceum movement, for example, succeeded because its blending of informal (discussion) and formal (lecture) structures matched well its hybrid populist and authoritarian origins. Mann's common school system had introduced reforms that were total in scope, thus "public," yet based on existing Academist structures (and bourgeois expectations) already present in private schools. Efforts later in the century to shift this system in the direction of the "scientific" agreed with these same fundamental patterns. But Dewey's point goes still further. It implies that the traditional school, extended by "scientific" organization, bears within itself the structural demand for recitation as a teaching method. The endless rote repeating of given material, that is, must be seen as prefigured in the teacher-centered classroom, where given authority is itself continually rehearsed in the smallest facts of interaction (the teacher stands, the students sit; the teacher lectures, the students respond; the teacher must control behavior and use a given curriculum, which the students obey and absorb; both the teacher and the students must produce "results" for routine evaluation). It was Dewey's feeling that the teacher had been made responsible for a system that utterly constrained what he or she could do. If his or her job came down, as it so often did (and still does), to keeping order first, this was in large part because he or she was fully subject to the same demand in a dozen different ways—indeed, the entire circumstance of schooling, especially when rendered "scientific" and "efficient," was based on the recitation of dependence.

This type of understanding helped urge Dewey toward an antiestablishment, Reformist position. In many ways, he remains among the greatest reformers of sensibility in American history—certainly since Jefferson, whom he resembled as well in his struggle and failure to reverse certain momenta of his time. His power was to change attitudes, especially at the level of feeling and hope. And yet, as experts, pedagogues, and parents, Americans have rarely opted for his ideas in practice, leaning instead toward conservative structures on the mass level. Dewey could not stop the progress of 19th-century mentalities into the 20th. In some ways, finally, he was too much a part of them himself.

Dewey was trained at the model for the new research university, Johns Hopkins (Ph.D., 1884), and accepted his first appointment at another, the University of Michigan. His study of psychology and philosophy was deeply influenced by the new scientific outlook. At the same time, he had spent time as a high school teacher earlier on and was acutely aware, personally and philosophically, of the vast changes and confusions that industrialism had introduced into American life since the Civil War. While at Michigan he was influenced by Hall's program for a curriculum more sensitive to childhood development, a concept that led him to join the Herbartians in the early 1890s, just before he moved on to the University of Chicago. Almost immediately, however, he found himself in conflict with the group's concept of "cultural epochs." If the recapitulation hypothesis were to hold, Dewey felt, it would have to be viewed differently, not in terms of literary "culture" but instead in terms of the evolution of human knowledge as it appeared in everyday social activity ("occupations" or "living knowledge").

In developing this idea, Dewey was less an originator than a synthesizer. Besides borrowing from Hall and the Herbartions, he absorbed direct influences from German romantic philosophy and pedagogy, and from teachers and mentors such as George Herbert Mead. His pragmatism, grounding all truth and knowledge in social situations and human conduct ("All truth is developed from purpose"), he took nearly whole from William James's *Principles of Psychology* (1890). His initial concepts about education, meanwhile, he developed in fuller fashion after going to Chicago in 1894 and there encountering the methods of Francis Wayland Parker, renowned reformer of the Quincy school system in Massachusetts, who had recently become principal of Chicago's own Cook County Normal School. In Parker, Dewey discovered his own gathering concepts *in vivo.* "We do not claim that nature is the center, neither do we claim that history and literature are the center, *we do claim that the child is the center,*" Parker remarked in 1895, in a speech before the Herbart group (Parker, 1895, p. 156). His most famous phrase— that the public school "should be a model home, a complete community and embryonic democracy"—was one that Dewey fast made his own (indeed, we associate it with him today) and repeated throughout his life.[2] Parker's method and approach, meanwhile, Dewey quickly adopted and modified as the basis for his famous Laboratory School (founded in 1896), where he put his ideas into practice.

Several pieces written in the late 1890s, "My Pedagogic Creed" (1897) and *The School and Society* (1899), established Dewey's reputation as leader of the liberal wing in Progressive school reform. Simply put, Dewey saw that new schools were needed for a new age: The Industrial Revolution had rendered the old normal and country–day schools, good for their time, obsolete in the modern urban world. The breakdown of the old way of family life, rural in essence (here Dewey, like others, tended to idealize the past), meant that children no longer witnessed the making of everyday things. They did not see trees transformed into firewood, burnt to make heat; they did not see vegetables raised from seed transformed into food for the table; they did not see wool shorn from the sheep, spun, and then knit into sweaters. "Home" as the place where things were planted, grown, cut, shaped, and refined into final social products no longer existed. The job of education, in some sense, was to put back this "home" into learning. *The School and Society* makes this idea explicit by posing the four questions Dewey thought of paramount importance: "(1) What can be done to bring the school into closer relation with the home and neighborhood life? (2) What can be done [to introduce] subject-matter in history and science and art that shall have a positive value . . . in the child's own life? (3) How can instruction in be carried on with everyday experience and occupation as the background? (4) How can adequate attention be paid to individual powers and needs?" The ideal school is therefore a place where the barriers between classroom and real life are dissolved.[3]

One sees that in Dewey the quest of education achieves a kind of maximal ambition. Not only should it replace the family, the lost home; it should now also substitute for—or in some way include and idealize—the neighborhood, the society, the nation as well. Like Parker, Dewey represents a recognition that "education" had become not simply a mass phenomenon but a massive collection of institutions, penetrating every facet of life. It had to be seen in terms of a larger whole, as a process of gradual formation in which families, peers, libraries, museums, newspapers, and much else took part. The school was anything but isolated; it was the nucleus of a centrifugal array of realities. But as such, it was where things could begin. To make society more democratic meant introducing democracy into the classroom, empowering the child as the active agent of his or her own learning. The school should be a place where society begins in its own best image, where children become "members," not just "recipients."

The fundamental image of "scientific thinking" in all this was tied not to nature or to functional organization, but rather to the desire for inspiring individual observation and knowledge building during a child's early years. Learning should be taught not by obedient memorization, but through "controlled experiments of experience," by explorations where the child learns to solve problems he or she develops. The frequent use of terms like "laboratory" and "experiment" suggests that Dewey saw the teacher as a facilitator, encouraging "investigation and discovery," not as an authority figure, demanding absorption and retrieval. More strikingly, he also agreed with Spencer on one other point: "The native and unspoiled attitude of childhood, marked by ardent curiosity, fertile imagination, and love of experimental inquiry, is near, very near, to the attitude of the scientific mind" (Dewey, 1910, p. iii).

In short, this "mind" for Dewey was near, very near, to the essence of learning itself. It was about continual improvement, continual striving after revision, a putting of every idea "to the test of consequences." Science for Dewey was an organic kind of knowledge that perceived the world in terms of shifting, dynamic unities, always seeking adjustment. The final metaphor it offered for learning was one that declined all technological aspect, that went beyond the biomechanical process of Mann, Spencer, or, in staged form, the Herbartians. Learning was, instead, continuous and unending "growth," with no final goal beyond itself—not a devised preparation for life but a form of living itself.

The appeal of such ideas, in an era characterized by prosperity and a search for conviction, came to be truly enormous. On the stage of national sensibilities, Dewey took over the role previously held by the grand Europeans. He was unique as the first American intellectual—the first (and in some ways, last) philosopher, especially—to move the domestic conscience so profoundly, with such provision for a discourse of conversion and idealism. Yet the gleam of his thought was not matched by the transformation of his ideas into institutional practice. In large part, this was due to certain important problems.

First, for all his concern with accessibility and demystification, Dewey wrote in a style that was frequently vague, elliptic, and difficult to pin down. This is particularly true for his most influential book, *The School and Society* (1899), based on three lectures he had made to parents and students of his Laboratory School. The oratorial quality of this work, its lyrical turns and grand generalizations, helped make it available for many purposes beyond those intended by its author. Beyond this, Dewey's output was huge, and many specific points made early on underwent significant change over the course of his career. Often, his reputation preceded all else, and the eagerness of the period for "new" ideas—especially those promising change and progress—encouraged many to read only isolated pieces of his work, and then to expedient purpose. Not a few of these ideas were lifted out of context and elevated to the rank of manifesto for various sects of Progressivism. Isolated and frequently bastardized, Dewey's ideas were used to promote "reforms" Dewey himself rejected: for example, those promoting pure vocational education for "poorer" students; stricter standards for teachers; a general emphasis on the school over everything else. In his later career Dewey spent much valuable time disassociating himself from such claims.

Second, Dewey stood firmly against the systematizing impulse which was by now central to most American notions of the need for teaching order, discipline, good habits, and moral virtue—values that had become synonymous with the idea of a good, productive, and (with regard to parental hopes) successful citizen. More in line with this impulse—or at least with the belief that the school should guarantee such values—were the behaviorist ideas of Edward L. Thorndike, Charles Judd, and Lewis Terman, who maintained, claiming a result of experimental science, that "the prime law in all human control is to get the man to make the desired response and to be satisfied thereby" (Thorndike, 1912, p. 96). This meant that "the great weapon of all who wish [change]—in industry, trade, government, religion or education—[is] to change men's responses, either by reinforcing old and adding new ones, or by getting rid of those that are undesirable" (p. 97). In actual practice, such theorizing meant total standardization, of teaching methods, administration, and student response. It was to be simplified and ensured by the use of numerical intelligence testing for purposes of ranking and tracking all pupils, and matching them against a hierarchy of curricula each of whose different levels would be designed to suit a specific range of innate abilities.

The appeal of such ideas called strongly upon desires, among industrialists and school administrators alike, to eliminate waste and promote efficiency in education. Thorndike and his followers advocated a new round of system-building that would essentially carry forward the reigning structures of teaching, organization, and accountability already in place. Their success was in part dependent on this promise of reform-by-streamlining which never threatened the existing order. Moreover, they extended their appeal in other directions, to old fears newly released from their bottles. Perhaps most of all, these programs had a direct hand in furthering nativist demands that the classroom (once again) act as "Americanizing" agents for the new wave of immigrant children that entered the schools between 1900 and 1915. "Scientific education" involved, first, testing these children to "prove" which of them had "lower ability" (naturally, the great majority), and second, the efficient teaching of habits appropriate to their level of possible achievement (the knowledge that would make them "the best citizens they can be"), such as cleanliness, politeness, some courses in the heroics of American history, family values, a trade or two. The effect, as Dewey more than once noted, was to systematically create a background psychology of inferiority, along with a general notion that becoming "American" meant the necessary abandonment of all foreign customs, languages, and traditions. Writers such as Cubberley made it clear, in no uncertain terms, that the school would help "assimilate and amalgamate these people as a part of our American race, and . . . implant in their children, so far as can be done, the Anglo-Saxon conception of righteousness, law and order, and popular government" (1909, p. 15).

To his practical detriment, Dewey stood vividly counter to everything that underlay this movement, from its behaviorist underpinnings to its views of Americanism. No more clear statement of this refutation can be found than in his laying out of the aims of a progressive education: "To imposition from above is opposed expression and cultivation of individuality; to external discipline is opposed free activity; to acquisition of isolated skills and techniques by drill, is opposed acquisition of them as means of attaining ends which make direct vital appeal; . . . to static aims and materials is opposed acquaintance with a changing world" (in Knight & Hall, 1951, p. 535). Such views were shared by other Reformists of the day, such as Wilhelm Wundt, Francis Parker, and Maria Montessori.[4] Dewey argued, too, against the reigning iconic theme of "the melting pot"—a technological image, one might note, fully in league with the "raw material" factory view of human beings (all original differences, that is, needing to be lost in the great and efficient smelter of American-ness). In place of this image, Dewey tried to promote a kind of multiculturalism, seeing, as Cremin (1988, p. 238) notes, "the need to redefine Americanism so that it would come to mean, not the abandonment of one identity in favor of another, but rather the combining or orchestrating of diverse identities."

In all these oppositions, then, John Dewey spoke as a kind of elevated conscience, but one unable to turn the tide of pragmatic action. The failure of Deweyan education as a model for the nation's schools was a result of its having arrived on the scene too late, historically speaking. The reforms for which Dewey struggled were just too radical, given what school had already become. The freedom he granted to the child and to the "embryonic community" he foresaw, while by no means total, was still a kind of utopian pastoralism, a protected stage of individual fulfillment in the midst of a social world whose wheels ground without end or mercy in the opposite direction. Dewey's followers thus found themselves in an awkward position: For those such as W. W. Kilpatrick, their job was less that of popularizers than of compromisers. They had to work very hard just to implement a fraction of Deweyan ideas possible with the given mass culture of schooling. Too often, their efforts ended in collusion with arguments for truncated or product-oriented learning. The effort, in any case, was not an enviable one.

Finally, Dewey shared the uncritical idealism about science of his time—an idealism that wholly overlapped that of his opponents. With regard to curriculum reform in this area itself, for example, he had this to say:

> Under present conditions . . . to be successful [the teaching of science] *has to be directed somewhere and somehow by the scientific expert—it is a case of applied science. This connection should determine its place in education.* It is not only that the occupations, the so-called manual or industrial work in the school, give the opportunity for the introduction of science which illuminates them, which makes them material, freighted with meaning, instead of being mere devices of hand and eye; but that the scientific insight thus gained becomes an indispensable instrument of free and active participation in modern social life. (1899, pp. 22–23; emphasis added)

The idea of "scientific insight" was never far removed from complete faith in the "scientific expert," in "applied science" as the means to egalitarianism. A mixture of naive hope and promiscuous assumption regarding science-as-savior tended to yield such confusions. The notion that "scientific thinking" could be a basis to alter not only education but the entire institutional structure of society toward full individualism and complete democracy was not merely idealistic, reminiscent of Enlightenment faith; it was self-annihilating. It could not but return to the elevation of "expertise," thus to inequality as the source of all potential equality.

As a spokesman for education, Dewey was at least as much a spokesman for "the scientific view," and for a type of "science" that merged only too well, in many confused shades, with that being used to argue for controls and standards very different from his own. In the minds of the public, school administrators, and politicians, the idea of "science" was often a muddle of different things: scientific method, laboratory work, hard facts, incomprehensible jargon, material progress, eternal laws, "the lessons" for acting, and, of course, teacher of systems and standards. Within such an environment, any praise of science as a fundament of answers in one area was bound to raise boats elsewhere. Dewey helped his antagonists, in part, by speaking the universal tongue of positivism. Though he would later come to view the use of science as often wrong and destructive, he never wavered in his faith that "the scientific spirit"—as a grand Academist movement of intelligence, purified of all possible bias—was the great transformational educator of the human world.

CHANGES IN SECONDARY EDUCATION

The changes in education that did take place in the early decades of the 20th century were varied. Ideas for making the school more responsive and responsible with regard to student needs were quickly translated into utilitarian Practicalist terms. A new perception that traditional classical study, as a reigning philosophy, needed to be altered, in terms of the formula that "the student was less a mind to be developed than a citizen to be trained," took hold. Behaviorist notions, translated into educational policy by those such as Franklin Bobbitt, and slightly later, W. W. Charters, argued that secondary school students were to be divided into distinct "learning groups," each of which was assigned an "appropriately fitted" curriculum. More academic patterns were retained for higher tracks, that is, "classical," "scientific," or "general" courses of study. Lower ranks were left with courses on the order of "household arts," "machine work," "drafting," and the like. Beyond this, school administration in many cities was greatly rationalized by various types of accountability, with each level having to report its "productivity," in standardized form, to its immediate superior.

The official shift in philosophy is revealed in the differences between two influential documents, published roughly twenty years apart by the National Education Association. The first of these, released in 1893, marked a sort of final summing up of the old traditional Academism. Indeed, in the same year that Joseph Rice published his study on waste and poor training in the schools, sounding the shot for a new era, a blue ribbon panel known as the NEA Committee of Ten headed by Harvard president Charles W. Eliot, came out with its own report demanding the very opposite of a search for innovative curricula. Faced with such realities as low attendance (many young people still opted for work rather than high school); increasing immigration; the fact that girls, whose career possibilities were very limited, comprised nearly half the total student body; and, finally, that the great majority of pupils, male or female, had no intention of going on to college anyway, the Committee felt obliged to come up with a universal policy that would best suit every life situation. What was their choice? Despite their open-eyed view of the general circumstances involved, this group of academic professionals came down heavily on the side of what they knew and what they honored most, a college-type scholastic program consisting of four separate programs: classical; Latin-scientific; modern languages; or English, with a core requirement that every student take at least four years of English, four of a foreign language, and three years each of history, mathematics, and (finally) science. The belief of the Committee was simple: Education meant gaining access to "high culture" and the "mental discipline" it contained. The result of such learning would presumably equip any student with the required intellectual and moral virtue to succeed at any profession, trade, or life-choice activity. The hierarchy of subjects was clear, too: The

old monarchical values attached to language and eloquence—incidentally, now given a new half-technical veil by
being relabeled "communicational abilities"—remained firmly in
place. Overall, the attitude of the Committee of Ten was that
the greatest good could be done for the youth of the nation by
forcing a full college curriculum downward into the high schools.
To the question "Should [a] subject be treated differently for
pupils who are going to college . . . to a scientific school . . . [or]
to neither?" (NEA, 1893, p. 6), the answer was a flat and
unanimous "No."

Within less than a decade, however, this type of belief and
its academic program had come under heavy fire. Such an
attitude, it was now felt, with its attendant curriculum as a
universal standard, might be fine for those few would actually go
on to college and seek academic or professional careers. But to
those destined for less intellectual work, for jobs in shops, the
manual trades, the factories, or the home, such scholarly study
was a waste of precious time—time that could be far better spent
learning skills and habits that would prove valuable, even necessary, to each of these types of existence. Education was to be a
progressive thing, giving access not merely to noble thoughts
and complex ideas but to the practical means of living, of
making a life in a changing, industrialized world. Thus by the
opening years of the new century did Progressive reformers begin
to claim the upper hand (ethically, as well as politically), asking
for diversification beyond the old collegiate core. A new document, issued by a new group, a Committee of Nine, finalized the
relevant philosophy in 1911 by stating that high schools must be
"adapted to the general needs, both intellectual and industrial,
of their students" (NEA, 1911, p. 559). Indeed, no more striking evidence for the change of attitudes involved could be found
than in the composition of the committee itself: In the place of
the venerable president of Harvard and his highly chosen circle
of academic professionals there now stood school superinten-
dants, teachers from vocational training institutes, principals,
and one professor of education. All of these participants expressed their belief in "an organic conception of education [that]
demands the early introduction of training for individual usefulness, thereby blending the liberal and the vocational" (p. 559).
The "bookish curricula" of the past had not only developed "false

ideals of culture," it had helped create a "chasm . . . between the producers of material wealth and the distributors and consumers thereof," and it had done this by "leading tens of thousands of boys and girls away from the pursuits for which they are adapted and in which they are needed" (p. 561). Classicism, in other words, was now demonized as orthodox policy. Usefulness was proclaimed the new guiding light.

The new philosophy took hold and swept aside all important opposition. Political battles between incumbent and newly-appointed administrators, between teachers, and within the NEA itself, were by no means absent or simple, but they were short-lived and conclusive. A skeletal version of the Committee of Ten curriculum was retained for the highest academic ranks, that is college-bound students (it has remained in place down to the present day), roughly 5% of the total at the time. All others were subjected to curricula designed in accordance with the type of "usefulness" to which their abilities would presumably best "adapt" them.

This idea of "usefulness," one should note, was not the same as Franklin's. It aimed not at a general openness to the world of labor and choice of work, but rather at a fitting-in of the individual into a given (even if changing) state of affairs. It was applied both to the student—his or her future role in society—and to the curriculum, which would act as a necessary shaping force for this predestined role. "Usefulness" was a philosophical and moral category, employed toward the hope of creating of students as nearly interchangeable parts in a well-ordered, decent, law-abiding, and productive society.

In the sciences, an interesting division came to exist with regard to the specific types of "usefulness" felt to be appropriate for each field. In the physical sciences, for example, educators believed that non-college bound students should be taught a few fundamental laws as a basis for solving or examining problems of immediate service to society. A rudimentary knowledge of forces or chemical ingredients would be used for a more lengthy study of how buildings are made or how a detergent works. Geologic principles would teach one about the importance of coal or ground water. Simple mathematics, meanwhile, were put toward computing the prices of everyday items, keeping a household budget in order, or the like. As a whole, the physical sciences were thus employed to teach students (in the lowest tracts especially) the facts of how society was put together, like a clock, and how one participated in this order on a short- and long-term basis. Biology, on the other hand, according to Pauly (1991),

> taught students how to think about the cosmos, what to eat and drink, how to behave toward life forms living and dead, what to consider clean and precious or dirty and dispensable, and how to react to intimate physiological functions in themselves and others. The ideal product of the biology course was a modern male—an individual whose physiological and intellectual development had converged, who understood his place in the world around him, and could act intelligently to improve it. (p. 663)

More than other sciences, then, biology was used to teach an organic morality essential to a good and proper social existence. More than physics or geology, it became a center of Progressivist ideals because of its ability to focus on the human subject and its self-activity, in public and in private. Political/moral contents for the making of a "better citizen" could be legitimized by being brought within the realm of the "scientific" itself: to be tidy and well-groomed, to observe proper dietary habits, to sanctify life, and to be restrained in sexual matters—these and other middle-class values were all taught as being "part of nature," necessary elements in "the order of life on earth."

Did all these changes yield new methods of teaching? One must return to Dewey's point about fundamental structure. Some new ideas in the classroom were tried out, certainly—for example, field trips, group learning, and individual laboratory work. Yet in spite of everything said about the need for "a new type of student," the new procedures implanted for ranking pupils, assessing teachers, and making administrators accountable ensured that no important departures would be made from the old form of instruction based on recitational forms of performance. True reforms in teaching method were largely confined to elementary grade levels, and then only in a relatively few cases beyond such experimental institutions like Dewey's own Laboratory School at the University of Chicago and Parker's Cook County Normal School. Also to be included here are a scattering of efforts by socialist reformers in the form of Sunday schools, local school cooperatives, and worker's educational societies, most of which were short-lived.[5] But, again, even in these more radical settings, new approaches to teaching and learning did not really last very long. Within less than a generation, by the 1930s, the notion of student participation was more doctrine than reality, with teacher-dictated lessons and lectures having become, or having remained all along, the practiced norm.

PROGRESSIVISM AND HIGHER EDUCATION: RISE OF THE EXPERT

Laurence Veysey (1965) has pointed to a major change in the structure of higher education during the last quarter of the 19th century and the first decade of the 20th. Nowhere else were these changes more evident and rapid than in the growing research universities and technical colleges (such as MIT), where, within the span of a single generation, a host of foundational "modernizations" took place that set the pace for academic organization thereafter. These changes included such things as "increasing presidential authority, bureaucratic procedures of many sorts, the new functions of the deanship, the appearance of the academic department with its recognized chairman, and the creation of a calculated scale of faculty rank" (p. 268).

Such changes directly mirrored those of most larger, urban institutions of the time. The rationalization of authority, its distribution into more complicated, presumably "efficient" hierarchical networks; the advent of ranking and certifying systems—all these became central to the expansion of large and rapidly growing social structures, from state legislatures to hospitals, from banks to schools. While there is no room here to discuss the deeper place and role of this movement within the history of capitalist organization, it does reflect a period in which authority within individual institutions became reconcentrated into a network of functional local tyrannies. Inside academia, as with the rest of schooling, this tyranny was justified by a new discourse often dependent upon the corporate work ethic; "efficiency" and "management" reigned here too, frequently enough. As a single example, the credit system was installed at this time as a means to quantify student productivity and eliminate "waste." Academist rhetoric about higher ideals, though still maintained, was now used more than ever to mask the realities of competition (for students, money, status), a rewards system, concern with student productivity, and the establishment of standards for degree certification.

Part of this process involved the beginning of faculty dominance, especially over curriculum decisions (what Christopher Jencks and David Riesman once called "the Academic Revolution" [Jencks & Riesman, 1968]). It was at this time, between about 1880 and 1910, when the sciences were completely filled out and divided into their present-day professional categories of "physics," "biology," "geology," and so on, each of which greatly expanded its total course offerings and created a comprehensive plan of requirements for majoring students. Total faculty control over content would not be secure for another generation, but by that time it was largely a foregone conclusion. The college, as a single unified entity, began to disappear. Knowledge and its responsibility came to rest with "the experts." And in science, as elsewhere, these experts were the literal and philosophic workers within major institutions. They were not simply the inhabitants or owners of their knowledge; they were its representatives, its popularizers, its dispensers.

In this atmosphere of change, purveyors of the elitist tradition required help from a new vocabulary. This they got, from "science." Compared to the time of the Lazzaroni and their struggle to get science generally accepted as a profession, the situation had entirely reversed: Positivism was now the standard, declaring that the method and logic of science could solve all "true" intellectual problems, and later social ones as well. Academic professionals and social activists reflected on how "the objective possibilities" released by training in science could be applied toward reform of all kinds. The historian Frederick Jackson Turner, for example, was quite explicit on this point:

By training in science, in law, politics, economics, and history the universities may supply from the ranks of democracy administrators, legislators, judges, and experts for commissioners *who shall disinterestedly and intelligently mediate between contending interests.* . . . It is surely time to develop such men, with the ideal of service to the State, who may *help to break the force of these collisions,* to find common grounds between the contestants and to possess the respect and confidence of all parties which are genuinely loyal to the best American ideals. The signs of such a development are already plain in the expert commissions of some States; in the increasing proportion of university men in legislatures; in the university men's influence in federal departments and commissions. It is hardly too much to say that the best hope of intelligent and principled progress . . . lies in the increasing influence of American universities. (1920, pp. 285–286)

Instead of the school, as with Dewey, the university becomes the source of all reform, the origin of both the knowledge and the trained capacity to fulfill "American ideals." Either directly, or as a model of "disinterest," science offered "progress" in a more humanistic form, as a type of consensus building, a means to "find common ground" for the betterment of society.

Yet Jeffersonian sentiments were but half the tale. Turner's call clearly places the idea of nonpartisan "service" in the hands of the new orthodoxy of expertise. Indeed, Turner was promoting what had already come to be known as "the Wisconsin idea" (an idea he helped found), based on the aggressive programs at the University of Wisconsin under President Charles Van Hise, whose vision of higher learning was that it be a source of applied science with the mission to better the lives of all citizens. This would be accomplished mainly by furthering research in all technical fields and by supplying experts to industry, government, agriculture, and society in general. As such, "the Wisconsin idea" marked a high point in Progressive beliefs about science. Indeed, as Cremin (1977) states, "In place of the self-instructed person of virtuous character . . . the Progressives foresaw the responsible and enlightened citizen informed by the detached and selfless expert, the two in a manifold and lifelong relationship that would involve every institution in every realm of human affairs" (p. 94). Reformist in spirit, this attitude was nonetheless based on an Academist view of science (removed and pure) which it sought to put to Practicalist ends. Shaking these various fluids of sensibility together produced a colloidal suspension of disbelief; science, "detached and selfless," could do no wrong.

Scholars imbued with Progressive ideals wanted, above all, to make their disciplines somehow useful to democracy, to equality. "Science" seemed to provide a way to do this, to yield both truth and the "lessons" to be derived from such truth for social betterment. Moreover, scientizing a field meant, symbolically and practically, its transformation into a profession within the university. The spectacular rise of this institution, accomplished within three short decades, had itself been founded on the ideas and imagery of "research." Institutionally, therefore, the attainment of full modern status meant either application of "the scientific" directly to a field, or else the uptake of related language and forms of publication.

Many fields—history, economics, philosophy, and psychology, for example—chose the former path, scientization. For most of these, this meant an evolution away from literature. No stable definition of "science" existed, as an intellectual approach: It could mean the collection of "facts"; quantification; an emotionless style; or a search for "theory." But in terms of basic discourse and professional practice, the patterns of imitation were nearly identical. One now spoke of "research" in scholarship, of "contributions" and "progress in the field." Moreover, the writing of books came to be replaced by journal articles, monographs, and "research studies" as the primary mode of textual labor, whose total yield now came to be used itself as a measure of professional worth. For fields like history, sociology, and psychology, it also meant the adoption of a flat, declarative style, prone to jargon, adamant against "literary technique."[6] Indeed, with its concern for the transfer of information over technique of expression, this documenary style of "factness" came to be taken up by newspapers, by government, by writing in nearly every area of official expression, from which it has profoundly influenced the direction of language change in 20th-century society.

In its literal and encompassing phase, the scientizing of the humanities enjoyed a short heyday. In the social sciences, of course, it carried on without pause; in philosophy too it continued (under the name of "analytic philosophy") well into the 1950s and 1960s. In history and literature, however, it either subsided into one of many "schools" or else disappeared altogether. Yet, by this time, the patterns had been set. The rise of the expert and his power within American society had already occurred, predicated on the early belief that all fields taking society and the human as a subject could develop their own fundamental laws and scientific truths. Not a few academics before World War I even predicted the complete disappearance of *Geisteswissenschaft* (humanities scholarship) under "the march of positive knowledge." As such, science and the scientific were more than essential territory for the playing out of social conflicts; they were the very agents of that conflict, the generators and furtherers of it—and not the least in education.

PROGRESSIVISM AND HIGHER EDUCATION:
CURRICULUM AND TEACHING

In the colleges and universities, meanwhile, the paean of "usefulness" was heard, but always applied selectively. The pre-eminence granted to research meant that scholarship, as an Academist mission, came first in any case. The aim was to create new knowledge first, the "useful citizen" last. Certainly, to guarantee further federal, industrial, and public support, vocational courses of many types were implemented, but always as low-status, segregated portions of the curriculum. Due to greatly expanding enrollments, which went from about 150,000 in 1900, to over 400,000 in 1910, to more than 1.2 million in 1922, all fields underwent a drastic phase of expansion—but especially the natural and social sciences.

The university, in short, heeded the demand by those such as Turner to become the center of expert training. More than willing, it was eager to be the forge for a scientific intelligence capable of taking up in objective fashion, dissecting, and solving all important problems of industrial society. This did not mean, however, that existing forms of Academist teaching were to be thrown out, especially in science itself. Here, rather, a scheme of division was employed to accommodate existing pressures for reform. After about 1915 course work in most universities and colleges became partitioned between a curriculum oriented toward future professionals and one for nonscientists. The first involved scientific teaching as it had basically come to exist, as a system of rigorous apprenticeship. Basic research and "original work" were strongly encouraged by means of the thesis system, imported from Germany after 1870. As Nathan Reingold (1991) has pointed out, the rise of the research universities brought with it a progressive increase in the amount of time required for the doctorate—from 1 or 2 years in the late 1880s, to a minimum of 2 years by 1900, to 3 years and more by 1916. The reason for this seems to have been a demand for greater amounts of high-level training in several fields, based partly on older Academist ideals that a "more rounded" education "was the necessary preliminary to both culture and utility" (Reingold, p. 187). But more critically, it reflected the growing attitude among academic scientists that "significant work" demanded more long-term and in-depth inquiry, and that knowledge of different specialties could fertilize such inquiry to yield research of the highest quality.

Other, more central aspects to Academist science came to exist, showing its partial derivation from the older classicism. Undergraduate textbooks, for example, now contained little potted histories of the "great ideas" and "great individuals" that had helped build each particular field. Lab work, in concert, changed from being a mere matter of demonstrations (whether by students or professors) to a kind of tutorial in which students repeated the famous experiments that had yielded the "great ideas" mentioned in their books. The "science" popularized by this type of teaching, in an era of mass education and growing competitiveness, was thus an inevitable and heroic march upward to the present, enacted by lone investigators employing the tools of genius. In working his way through a particular course, the student commonly recapitulated this simplified and purified "science," which thus remained a coveted professional image for decades to come.

The other part of the curriculum involved the creation of a "general science" course for nonmajors. Here, rather than (or, at times, in addition to) a succession of great ideas, scientific subjects were taught as a series of important, everyday principles and moral maxims regarding the value of "observation," "evidence," "cooperative activity," and the like. The laboratory was similarly reduced to providing the student with exposure to the "wonders" of ordinary life, encompassing everything from "kitchen chemistry" to "industrial arts." As time went on, into the 1930s and 1940s, this eventually gave way to the more familiar survey course, in which the major concepts, discoverers, and experiments were taught in a paradelike style (still Academist), usually supported by "gee-whiz" lab work. Whether in 1920 or 1950, however, academics tended to view such courses as a burden, a "stooping to the masses"; more often than not they were given to young, incoming assistant professors who had the least teaching experience and the most demands with regard to achieving career security. All along, the unspoken desire was that this type of course, confined mainly to the freshman year, descend into the high schools. This it eventually did in any case, during the 1950s, where it became still more simplified. The point, however, is clear: "General science" was used as an argument for not studying professional science (as learned and practiced by scientists). The division of the curriculum, therefore, had its own Academist function, for it sealed off in removed and elevated fashion this science of real-world research.

Such a division was obviously not what Dewey and his followers had in mind. Academic science and engineering were both shielded from any external criticism or deep change by this new "science for the citizen" approach. The Reformist cause, with its own uptake of science as savior, was hardly in any position to challenge technical learning at the highest level. Even philosophically, it rarely examined such things as the character of science as a profession, how it fit in to modern society, what its historical role was, or what were its institutional connections and demands. "Science" was something far too grand and pure for such things, and yet, far too helpful for showing how the individual could adjust to, find both wonder and improvement in, the given environment.

Beyond these things, the Progressive argument for "individual usefulness," was sooner or later destined to help support what Noble (1977) has called "the corporate reform movement in American education," by which universities came to "direct the process of scientific research and to create an educational apparatus which would meet the demand for research manpower" (p. 168) so central to the expanding corporate industrial system. In the name of efficiency and of increasing the prospects for Americans of all levels in the economy, industry increased its donations of money, equipment, and employment possibilities to the universities, and thus helped directly set research agendas. That it could do this while employing so readily the Reformist rhetoric of the day testifies again to the open-ended and confused character of this discourse, its reliance on vague invocations of "public service," "education for living," and "science for society"—demands that were only too easily taken up by power brokers for the given order and modified to serve economic advantage and blatant nationalism.

With their unblinking faith in scientific thinking and truth, the Reformists of the early 20th century never thought to question the foundations of their own belief—nor, for the most part, the status and methods of the men who represented it professionally in the universities. In the end, Cubberley therefore found his ready echo in higher learning through the language of those such as Henry Suzzalo:

> The American system of schools has a sanction in public efficiency as well as in equality of personal opportunity. It is a special system of getting brains for the public purpose. University [educators] have an immediate responsibility to make the prospect more effective. . . . Soon we must become as wise in pedagogical method as we have long been in scientific method. The processing of human beings through intellectual experiences is far more important socially than the processing of material things. (in Noble, 1977, p. 224)

What Bearing It May Have: Legacies of Progressivism in the Early 20th Century

By the early 20th century the most ancient and hallowed forum for learning in the Western world was in eclipse. No more would the tutor–pupil relationship, with its bonds of authoritarian intimacy, govern cultural practice with regard to education. Where this relationship existed at all, it was mostly held in reserve for the highest orders of study, the graduate school. Everywhere else the teacher as lecturer ruled, preacher of knowledge to ever-growing, ever-more diverse congregations of students. The reign of mass education, in short, begun earlier for the elementary grades, now encompassed all of secondary schooling. By the 1920s and 1930s the influx of new pupils into the high schools was unprecedented, aided by immigration, by the enforcement of child labor laws, and by the raising of the minimum age for leaving school to sixteen. Enrollment rose as a result from about 800,000 in 1900 to over 2.5 million in 1920, and to more than 4.8 million a decade later, a fivefold increase in a single generation.

To many, school was now a form of "social service," the first great element of what would later come to be known as "the welfare state." What William Boyd (1966) pointed out for the 1950s was no less true even 20 years earlier—that, in fascist and democratic nations alike, "any school's chances of viability for itself [now] depend on its being geared into the total planning of statecraft—no matter what the politics of the government" (p. 470). Over and beyond the hopeful discourse of John Dewey and others, this massing of students and resources, as well as the basic rationalized and standardized structures into which they were arrayed and by which they were measured was the major legacy of the Progressive era. And the advent of scientific curriculum making in the 1880s and 1890s was the clear historical sign that this would happen. From the very beginning, "educational science" meant forms of standardization, and these in turn meant a deindividualizing of the teacher–student relationship.

Tuned by "social efficiency" and by ideas of "the scientific," mass education on the secondary level denied the value of the old classical learning, especially at the college level, even more than the sciences themselves had. Not only did such an education seem to erase the higher individual; it bore losses for the old Academism on philosophical, psychological, and political levels as well. With the coming of mass schooling—and its complement, mass culture (movies, dime novels, and radio)—a tendency grew up among many humanists to perceive a dilution in, or an alienation of, the personally sacred character the "great works" might have. Education on such a scale meant, finally, a loss of privacy and counselorship. Progressive schools, intended to give dignity to the masses, signaled the irrelevance or isolation of such piety. What this meant in turn (for such attempts at legitimacy were never far behind) was a privation of moral example, a decline in the opportunities for greatness, both with regard to individuals and American leadership as well.

The recurrence of traditional Academism, then, might be seen itself as a sign. Ever since the Condemnation of 1277, in fact, when the new "pagan" literature of Aristotle was forbidden in the University of Paris due to its conflict with Holy Writ, the sacred nature of "great books" has been a stage for projecting the lost or declining power of Academist intellectuals generally in times when new forms of learning emerged. The cyclical call to arms for the "liberal arts" is symptomatic of a period when education undergoes strong and rapid popularization, along diverse lines and at many levels.

But the advent of mass schooling was only half the story during this period. For the avid campaign in favor of the old form of undergraduate Academism, launched in the 1930s by Robert M. Hutchins, president of the University of Chicago, also saw itself as a campaign to save American democracy as perhaps the last preserve of Western values and culture. In an era when totalitarian regimes were rising on all sides, the study of the "great books" seemed to some an act of cultural redemption. Hutchins and his followers rarely put things in such terms. They tended to speak only in grand generalities, as befitted their program for a return to classicism. But lurking within their call for stability, for an "aristocracy for all," was a plea for ethical rigor and predictability they hoped would strengthen the country, or at least its future leaders. Dewey perceived in Hutchins's program its ironic self-defeat: "I would not intimate that the author has any sympathy with fascism. But basically his idea as to the proper course to be taken is akin to the distrust of freedom and consequent appeal to *some* fixed authority that is now overrunning the world" (1936, p. 104). Yet, by the late 1930s, while the high schools continued largely as they were, a core curriculum was being reestablished in most colleges, and academic politics shifted distinctly to the right (see Schrecker, 1986).

At the same time, however, one finds that humanists were not the only ones to perceive a need for moral uplift, in the Great Depression era. Scientists too, and a host of writers for the public, felt a strong urge to invigorate and guide ethical attitudes for the "democratic way of life" during a period when Hitler and Stalin seemed to threaten the entire world. Mass education was perhaps less the point here than attempts to popularize science for the average citizen, once again as a source both of national power and of virtue (personal and public), the latter now helping to mitigate any excesses of the former. This tendency was not a movement, properly constituted. It had no organization, no designated leaders. Nor was it confined to science alone, or expressed in books intended only for well-educated readers. Rather, it was loose and broad-based and therefore can be taken as an indicator of a widespread impulse to instruct the public about a path to better living, higher conduct, and more patriotic citizenship. As such, it was Progressive in outlook. It consisted of books that, while not aspiring to the ranks of the "great," individually sought to reestablish the tutor–pupil relationship, and to do this through the "teachings of science."

More generally, the period of the 1920s and 1930s divides itself, culturally and educationally, on either side of the Great Depression. Throughout this period positivist ideas continued to direct a great deal of pedagogical thought. The early to mid-1920s, for example, saw the social efficiency movement reach its height in several areas: in vocational training (greatly advanced in the wake of the Smith-Hughes Act of 1917, which established yearly funding for each state for vocational programs in the high schools); in the increasing use of standardized aptitude exams; in the wide use of "output"-oriented curricula; and in the pedagogic theories for "education engineering" of men such as Franklin Bobbitt, David Snedden, and W. W. Charters.

THE 1920s: PROSPERITY AND POSTIVISM

Spared the serious ravages Europe suffered during World War I, the victorious United States now emerged as a new world power, fully conscious of the fact. The early postwar era, moreover, continued the trends of economic advancement begun after 1900. Corporate wealth expanded. Financial institutions grow bigger and richer, aided in no small part by technological development in general and by the great collection of specific industries that flourished in the wake of the automobile: oil, rubber, steel, chemicals, electronics, tourism, construction, to name but a few. The number of newspapers, magazines, and other periodicals reached a U.S. apogee between 1914 and 1930, while the movies, which only a few years earlier had been a drop-in activity for the working classes, now became a massive industry based on star worship and drew weekly attendances of up to 80 million by 1929. Radio, too, came on the scene and acquired mass audiences in the tens of millions before 1930. Images of material progress, of national confidence, enjoyed a weedy growth in all these media, and the sins of laissez-faire capitalism, subject to such intense scrutiny only a decade earlier, were now often forgotten or overlooked.

The early post-war period was also "the take off point in the history of American higher education" (Levine, 1986, p. 38). Colleges and universities that had lost enrollment during the war years now expanded rapidly, often by 50% or more before 1920 and as much as 10% a year through the 1920s. Schools were urged to offer a large array of new subjects, all of them in professional areas such as business, journalism, and engineering. Industry and government both provided funds for setting up new departments, establishing professorships, laboratories, work-study programs, and the like. The lion's share of this new support went to the sciences. Engineering enrollment alone grew from roughly 30,000 to 50,000 in the first few years after the war ended. As David Levine writes:

> By responding so quickly and so confidently to the public demand for practical higher education, American universities contrasted sharply with their European forebears. Visitors crossed the Atlantic Ocean to examine the kind of university that American energy, democracy, prosperity, and pragmatism had wrought. All were amazed by the magnitude and quality of America's collegiate facilities. During and after World War I, research laboratories and libraries were being built in the United States on a scale envired in contemporary [Europe]. The vice-chancellor of the University of London was shocked that Catholic University of America had as fully equipped and nearly as large a chemistry laboratory as any in Great Britain. . . . (1986, p. 41)

The war, then, had powered the United States into a new era of prosperity and self-confidence. And at the root of this confidence lay a gaining belief in the benefits of professional training at the college level, source for a new generation of leaders.

Positivism, as a general outlook, as a general faith in the benefits of "science," reached its absolute height during this period. Though certain fields, such as history, had by now turned away from literal analogies, others such as sociology, economics, political science, and psychology had become still more theoretical and quantitative. On the popular level, meanwhile, "science," "invention," and "progress" were terms used more than ever interchangeably, especially in the mass media.

Looking back but a few years, Frederick Lewis Allen wrote of the 1920: "The prestige of science at this time was colossal. The man in the street and the woman in the kitchen, confronted on every hand with new machines and devices which they owed to the laboratory, were ready to believe that science could accomplish almost anything, and they were being deluged with scientific information and theory. The newspapers were giving columns of space to inform them of the latest discoveries . . . intelligence-testers invaded the schools in quest of I.Q.'s . . . The word science had become a shibboleth" (1931, pp. 164–166).

Of course, what Allen and "the common man" usually meant by "science" was technology, or rather science as producer of new "machines and devices." But basic research was also beginning to be viewed by some image brokers as a source of public virtue. Indeed, the founding of modern science journalism at this time by newspaper publisher E. W. Scripps, who created the first syndicated wire service devoted entirely to technical news (Science Service, organized in 1921), had the stated purpose of promoting "the importance of scientific research to the prosperity of the nation and as a guide to sound thinking and living" (in Nelkin, 1987, p. 89). Scripps's project, decidedly Progressive in its ambition, signaled the early stages of a new era of science popularization.

The desire to repopularize science, in fact, had two basic sources. First, there was a growing perception among journalists, policymakers, and writers that, despite its real-world power, scientific knowledge was more distant and inaccessible than ever, that a greatly thickened barrier now separated the uninitiated layman and the initiated expert. This feeling, itself a legacy of Progressive attitudes regarding democracy for all, was heightened by the second source of popularization: lobbying on the part of scientists and national science organizations for more public support of fundamental research. The problem here was that both government and industry tended to fund research of a more utilitarian kind, with a clear view to introducing new technology. A growing number of scientists began to foresee the enslavement of their "higher" work to commercial demands (in the 1930s, this perception would help drive many scientists to join the Communist party, of which they probably came to comprise roughly half the total academic membership in the United States; see Schrecker, 1986, p. 44). Efforts were made to garner private contributions under the purist ideal of "no-strings support" and, at the same time, the somewhat less pure objective centered on "the health of American science."

These two popularizing trends arose from opposing desires. Where one aimed at making science accessible "to the people and for the people," the other hoped to further the cause of basic research, the very source of "inaccessible" science. Both, however, had a vested interest in creating a public taste for scientific knowledge. Newspapers clearly believed they had found a new form of news capital, a marketable product to be sold via appeals to Gettysburg Address democracy. Scientists, on the other hand, found themselves supporting the same types of images for their purpose, and thus went well beyond the limits for selling science set by Bache and Henry more than 50 years earlier.

The reign of positivism and the idea of "science for a better tomorrow" continued as well in the educational arena. In a way, the Reformist movement did bring new life to older, Jeffersonian ideals with regard to the value of scientific education. Yet notions of how such education might itself contribute to the creation of a true communal democracy now regularly entwined themselves with the economic and nationalist rationale that had been reinvigorated from the gilded 1870s and 1880s and applied to the "platinum 1920s." The major debate in higher education between the world wars, for example, was over the curriculum, and involved a return to the old battle between vocationalism— now risen to the status of professional training for high level careers (scientists, engineers, economists, businessmen, and the like)—and classicism, in its purist form. Colleges and universities tried to satisfy both demands, mutually exclusive as they were. Survey courses—known as "general education"—with a more or less classical bent, were instituted for the first two years, while honors programs, also begun at this time, carried the philosophy of "higher culture" into the junior and senior year. These innovations, however, pale beside the veritable explosion in professional departments and programs offering study preparatory for graduate-level training in specific career areas. At best, the hope was to produce practical scholars and scholarly businessmen—"idealistic materialists" in David Levine's words (1986, p. 93).

Progressive educators, meanwhile, argued vehemently against influential conservatives such as Robert M. Hutchins and Abraham Flexner. Such men described the new universities in terms of mass consumership, as "department store" or "service station" schools. To the Progressives, higher education was merely—or gloriously—responding (at last) to the demands of society itself: The university was finally becoming a central, interactive institution in modern American life. Yet there was something, too, in their rhetoric and philosophy that gave an unfortunate warrant to some of Hutchins' and Flexner's discourse. Though Progressive reformers of the 1910s and 1920s, such as Charles R. Van Hise, William Kilpatrick, and Walter Lippmann, may have had little if any direct interest in the "technological sublime," they nonetheless often spoke of the college as a type of factory, productive of "manpower resources" or future leaders in a "better mold." Indeed, it might fairly be said that intoxications of the technical, mechanistic type did not at all diminish. On the contrary, the burden of faith in the larger transformational power of the "scientific," whether as a background image for conceiving how society and its institutions worked or as a model for knowledge, continued to ascend still further.[1]

Walter Lippmann, for example, took up the idea of "the great society" and placed his belief for change in political institutions guided by trained technical expertise. His proposal was entirely a positivist one: to set up "intelligence sections" within the federal government filled by "disinterested, tenured experts," committed only to the search for truth. These experts would "investigate" the deeper workings of society—which, during this period, tended to be spoken of purely in mechanical or chemical terms, as a vast array of interacting forces, balanced or imbalanced, set in motion by catalysts or reactions—and the findings from their inquiry would then be turned over to policymakers directly, thus "bring[ing] to light the hidden facts . . . on which men can act" (1922, p. 29).[2]

At the same time that science was being popularized for public consumption, the "scientific" was being offered as a justification for removed and nonaccountable policymaking. Science for an informed democracy had its interesting correlative in "the scientific expert" for plutocratic leadership. Lippmann's "intelligence sections," of course, were nothing other than a version of the research university itself, source of light and objective wisdom, whose implanting within government would inject enlightenment into politics.

Similar beliefs in "scientific objectivity" and in the larger "lessons" to be gained from technical methods were widespread elsewhere. Indeed, during the 1920s they came to be shared by an enormous range of groups interested in social change. Many socialists and Marxists, for example, felt that a true "scientific view" of capitalist society was required to reveal the machinelike workings of the class structure, and that better instruction in the schools should include the learning of scientific principles and analysis of how these processes were reproduced on almost every level. On the far right, meanwhile, other "truths" had their own appeal. The glare of social Darwinism had dimmed, but only to be lit more brightly by the importation of eugenics from Britain (the American Eugenics Society was formed in 1926). Here, the class structure received a different type of "scientific" imprimatur than that offered by Marxism. Making use of the work of Louis Terman and IQ testing, eugenicists in the United States argued strenuously for a need to "rebalance differential rates of reproduction" among the various classes. The poorly educated "races," which tests showed to be where they should be, were reproducing at far greater rates than were the superior individuals of society.[3] "Science" was employed to argue the type of problems that would arise from "racial mixing"—problems that strongly argued the need for ability grouping in the schools. Schooling itself was viewed in terms of its revelation of "achievement levels," therefore as an "objective" measure of innate human worth.

Eugenics was held in contempt by most Progressives. Yet, as I have already noted, the reforms they favored were not wholly free of overlapping assumptions. One central idea concerned a

broadened curriculum, one more open to the needs of each
individual and society as a whole. This derived from the perception that most students did not go on to college, and therefore
needed an education that would prepare them for more "ordinary" lives and for "good living." The curriculum had to adjust
itself, differentiate itself, take in the realities of social life. These
"realities," however, were often thought to reflect fundamental
abilities: Standardized testing had helped "prove" what many
Progressives (though not Dewey) actually suspected, namely,
that college was not for everyone; that there existed different
types or categories of students; and that these types could be
"sorted" according to basic intellectual competency. Some students might be destined for "brain work," but most were headed
for office work, industrial jobs, or manual labor. A combination
of natural endowment and social background destined this type
of sorting, but the finality of the result was not very different
from that assumed by the eugenicists.

Progressivism therefore tended to ennoble labor as the aim
of schooling, and schooling itself as "preparation for living."
Hopes for reforming education, and perhaps with it American
democracy, ended up being guided by the concept of "adjusting"
students to their proper (inevitable) level of social achievement.
What this did, in turn, through such things as tracking and
curriculum programming, was to transform the school into a
detailed replica of the existing order of inequality in society. The
document of orthodoxy for this program appeared in 1918, as the
famous *Seven Cardinal Principles of Secondary Education,* released
by the National Education Association only seven years after its
Committee of Nine report. In brief, this document fine-tuned
the earlier philosophy, laying out a core curriculum of study
areas: health, fundamental processes, worthy home membership,
vocational efficiency, civic participation, worthwhile use of
lesiure time, and ethical character. The term "fundamental
processes," which stood for the traditional academic subjects,
indicates that the old psychology of faculties was still floating
around to some degree.

Otherwise, what is notable about this list is its attempt to
encompass the entire individual, not merely his or her intellect.
Nearly every area of human activity, from home to work to play

to personal attitude, was to be contained and shaped by the curriculum, which therefore seemed not so much a "preparation for living" as a preparation *of* living. To the degree that it was a place for the "useful arts," the school was a training ground in the art of being useful. Progressive education sought to give dignity and opportunity to the lower classes by offering their "life" a large and direct place in the curriculum. Yet this effort was never really separate from the idea that these people were incapable of more, that only a certain few could ever attain higher levels of interest and achievement. A Reformist spirit looked to Practicalist ends, and in the process yielded itself to an Academist plan.

In the 1920s this curricular scheme was further rationalized by influential educators such as Snedden, Bobbitt, and Charters, who gave it a "scientific" mask. These writers had taken the management ideas of Taylor and the behaviorism of Thorndike and elaborated them for curriculum building. Their main premise was that no such thing as general skills existed; each area of study developed its own, specific ability that could not be transferred to other areas. In Snedden's convoluted wording, "every distinguishable species of education and, of each species, each distinguishable degree, is or should be designed 'to meet a need' " (in Kliebard, 1992, p. 45). The biological metaphor is telling; it seems aimed at naturalizing what is otherwise an instrumental, mechanistic program. Each "species," for example, should be blueprinted to fill a specific social role for the "educand" (Snedden's word for student). For Charters, this role could be precisely defined by means of something called "activity analysis," in which the actions specific to a particular job or occupation could be identified, described, broken down, and quantified. Finally, such analysis, according to Bobbitt, could then be applied to a list of agreed-upon objectives, which would help dictate those activities and skills (mental or manual) that needed to be emphasized, refined, or otherwise strengthened through the curriculum.

In reviewing these ideas, Kliebard (1992) points out that the basic logic of design has remained standard for curriculum building ever since. This design involves selecting goals (for behavior); "operationalizing" them into distinct skills or activities; and deciding what subjects and what kinds of teaching will best develop such skills. The rationale behind this approach is a model of bureaucratic projection: It inscribes a system of curriculum specialists, each assigned a particular task, each responsible in a hierarchy of authority reaching up from the teacher at the bottom (he or she who must receive and apply the decisions of others), to those who decide upon the basic goals at the top. At the time, it fulfilled the desires of a range of interests—the efficiency movement, the vocational ed movement, the teacher certification movement. The result was that the high school curriculum in particular, but to a certain degree college programs as well, became even more differentiated, tracked, and ability-grouped. Courses of study now often came to include such things as machine work, household arts, industrial arts, carpentry, with subareas under most of these.

At the university level, meanwhile, such ability grouping took different forms. Beginning at Harvard in the 1920s, honors programs were implemented to encourage and divide off students considered to be superior. At the same time, simpler versions of many standard courses—for example, English for business majors or physics for humanists—were also added or proposed. More broadly, a large percentage of colleges and universities enacted a separation of the undergraduate curriculum into a "junior college," encompassing the first 2 years, during which a required "general education" course was pursued, and a "senior college" for the pursuit of upper level, professional, or advanced academic course work. This division was a result of several immediate perceptions: first, that many students did not complete more than 2 years of college work and thus required some acquaintance with traditional higher learning; second, that, despite this fact, too many students of mediocre talent were still going on to the junior and senior years, and so some type of gatekeeping mechanism had to be put in place (required exams or minimum grade average for entrance to the senior college) to keep their numbers down. As Levine (1986, p. 100) describes it, through such measures "a school or faculty could ferret out unwanted students [and thus] draw the line on mass higher education without risking the social and political consequences of limiting enrollment to the university as a whole." The "junior college," in a sense, was partly a concession to the classicists, since "general education" often meant year-long survey courses in the "great ideas" mold. The "senior college," then, represented an agreement to the growing demand for professional training in business, engineering, the social and natural sciences.

The curriculum thus became the object of specialization and a tool for partitioning presumed abilities and skills no less in college than in high school. The overall effect, in a matter of speaking, was to merge the pedagogic theory of those such as Bobbitt and Charters with actual educational practice. Especially at the secondary school level, it proved a merger that favored a new type of highly tuned, technological model for producing the most correctly "adjusted" student. It was a model whose logic of interest fell more on what a student might possibly do, how he or she might labor and behave, not who they might be or might become. It was concerned with people as participants, as actors or ingredients in a well-ordered, decent, working national society. But what this meant, too, was a disinterest in them as human beings, something that effectively negated almost every honest desire for their advancement. Lost in this model, after all, were both student and teacher: The whole burden of effort came down to controlling and dictating, transforming into "technique," as many aspects of the interaction between these two entities as possible. It was an educational philosophy that sought design of the individual first and last. The old republican values of civic and moral virtue, of discipline and character, had become designated categories in the curriculum itself, headings under which the dictating of social attitudes and behaviors might take place. Science would not lead the individual to these values as Jefferson had proclaimed; instead, it would design the individual to follow their contour.

The business analogy, meanwhile—the school as a factory—had its own specific meaning: It revealed that mass education, in an era of industrial preeminence, had come to produce learning itself as a form of work, a labor that obeyed demands for "product." This was obviously not the work that Dewey envisioned; yet it was work that reflected the outlook of its time.

The 1930s: Constancy and Change

As with many other aspects of American culture, the Great Depression brought change with regard to attitudes about science and education but constancy too.

One strain of thought blamed science for what had happened (see Allen, 1992; Kuznick, 1987). Decades of science promotion in the name of national and industrial progress made this a likely outcome when such progress stalled and finally came to a dead halt. Emotions in the South and elsewhere still smoldered over the outcome of the Scopes trial. Intellectuals were beginning to grow wary of expertise as a panacea, and more than a few writers saw moral weakness in the public's previous adulation of things scientific. More than a few were now ready to proclaim science among the greatest evils threatening humankind: "Scientists more than any other single factor," wrote one sociologist of the day, "threaten the persistence of Western culture. . . . Their calling has become a cult, a dark mystery cult. . . . The 'pure' scientist has to be a moral eunuch or a civic hermit" (in Kuznick, 1987, p. 39).

Such hostility, however, still was characteristic of only a minority. During the years of the depression, science was more often viewed either with qualified trust or with optimism regarding its ability to shape a better future. Much of the New Deal was strictly pragmatic, focused on creating more jobs as quickly as possible. Technology, source of "the machine age," was frequently blamed for automating many workplaces and costing men their employment (the success of Chaplin's *Modern Times*, released in 1936, testifies to widespread feelings along these lines).

There was a perception on the part of educators that the public had lost faith in technical knowledge and its professions: Though college enrollment continued to climb slowly upward, the number of students choosing fields in the sciences and engineering declined overall. This, however, was more a result of the changed job market, related to the downturn in industry, thus industrial research and development. (Companies decreased or stopped altogether the campus recruiting they had so actively pursued during the 1920s.) At the same time, the New

Dealers did not turn away from promoting "higher skills" in education. They were not foolish enough to place their faith in a return to a pastoral or craft economy; indeed, they saw that productivity depended upon well-trained personnel. To help support university enrollments, the federal government, through its Federal Emergency Relief Administration, enacted a major work-study program in 1934. Over a period of 8 years, more than $93 million was spent to aid more than 620,000 students (Levine, 1986, pp. 197–198).

From 1933 to 1941, moreover, New Dealers also underwrote a new child study movement. This they did by supporting a large network of nursery schools across the country (nearly 1500 by 1937), with a far broader spectrum of curriculum activities than had ever previously existed for such schools, including storytelling, drawing, playing with blocks and nature materials, learning hygienic habits, and (of course) learning proper social behavior. The motive was to help care for the children of unemployed workers, to provide jobs for teachers, and to help in parent education. But the effects were broader. For the first time preschool children were given a true curriculum, one that took its cue—and in turn helped nurture—the budding field of developmental psychology, which had offered a new kind of evolutionism through such ideas as "maturation" (Arnold Gesell) and "cognitive stages" (Jean Piaget). As a result of Works Progess Administration (WPA) this area of science, a direct descendant of child-study, was used to bring the nursery school to the borders of the existing mass education system, and to begin the transformation of children aged 6 and under into "students" (although this term is never used, even today).

The government's sometime mixed attitude toward science was not shared by others. During the early 1930s, Lippman's idea of "tenured experts" was reconstituted in the form of the Technical Alliance or Technocrat movement, which enjoyed a brief heyday between 1932 and 1934. Basically a Marxist-inspired enterprise, the Alliance was formed by engineers Guido Marx, Walter Rautenstrauch, and Howard Scott (impresario of the movement), along with critic and economist Thorstein Veblen. Its primary proposal was to establish "a corps of engineers and technicians who would run the economy in accord with sound engineering principles designed to maximize production" (in Segal, 1985, p. 47). For several years the Technical Alliance attracted both public and government attention, but it soon fizzled due to internal dissent and growing external criticism of communism. Yet the movement's initial popularity reflected the continuity of hopes placed in "objective science" as a means to solve social problems for all.[4]

Technocracy had been based on Taylor's "scientific management" ideas, which were viewed as the sacred path by which efficiency could be injected into a foundering industrial system. Yet by this time Taylorism was losing its place as the premier model for all managerial practice.[5] By the mid-1930s the depression had actually deepened, and a new moral scheme, oriented to provide hope for the individual, began to gain favor. Existing ideas of "mass production" and "educational engineering" were increasingly viewed as demeaning, as a poor and unfit ideology for inspiring belief in the future.

The hesitant attitude much of the public (and mass media) seemed to feel toward large-scale forms of social organization, toward big industry for example, helped spawn a desire for styles of discourse that made their appeal on a more personal, emotional level, that might promise general betterment through improving such things as moral fibre, personal loyalty, and honesty, especially for those in positions of power. It was at this time, in fact, when such feelings helped inspire a new type of popularizing movement in science. Numerous writers in many fields began to adopt science as an ideal for showing the public how to be better human beings: how to become ethical lawyers, more informed politicians, smarter parents, more honest bureaucrats, and so on—all with the idea of advancing democratic norms. Science would be a teacher to the average citizen, a personal teacher of honorable individualism. Yet there was no little confusion about exactly what "science" meant, and therefore what it could offer. Such confusion, for example, is apparent in the many (groping) attempts to try and provide a terminology that might aptly describe the very special state of mind that science was capable of teaching. Between 1930 and 1950 a host of phrases were adopted or inaugurated and for the most part abandoned: the scientific spirit, the scientific attitude, scientific morale, scientific honesty, scientific discipline, the scientific mind, and scientific character.

Such terms found especially avid use during the 1930s and early 1940s. Each was supposed to signify some essential feature of what could be learned from a study of science. The very scope of their combined intentions—which included mind, spirit, personality, and behavior—reveals how broad and deep such study was presumed to go. This belief focused on "science" as an ultimate source of all rational thinking and acting. There were Jeffersonian and Deweyan alignments: In the teachings of science one would find the very embodiment of democracy and therefore a democratic education—a "mind" gaining in character and probity as it informed itself about some problem in the world, then acted in accordance with this new knowledge. But on the other side, there was also a tendency to return to "science" either as the messiah to solve all human problems or as the vehicle of mythic national destiny.

Most books written at the time extolled the personal benefits of knowledge. These included "skills" for life, for employment, and for freedom. Science was at the top of the list, praised for its power to culminate human ability:

> The scientist is a man of integrity and faith who trusts the basic laws of nature and intelligence to lead him into the paths of truth. His loyalty to truth is unquestioned: his capacity for patient and sacrificial inquiry is limited only by his powers of endurance; . . . his objective is the welfare of mankind; and his discoveries . . . are the free possession of democratic peoples. (Cole, 1938, p. 23)

Faith in the nonscientific benefits to be gained from scientific knowledge came to be shared by writers in nearly every field, whether they believed in "scientizing" the humanities or not. Here, for example, is a passage from Jerome Frank's enormously popular and influential book *Law and the Modern Mind* (1935):

> While lawyers would do well, to be sure, to learn scientific logic from the expositors of scientific method, it is far more important that they catch *the spirit of the creative scientist,* which yearns not for safety but

risk, not for certainty but adventure, which thrives on experimentation, invention and novelty and not on nostalgia for the absolute. (p.
98)

Like Frank, liberal reformers of the 1930s and 1940s often
emphasized this aspect of "adventure" and bold skepticism—long
adopted by scientists and science writers themselves as a kind of
self-elevating trope. To such reformers, the creative scientist was
not merely superior; he was superhuman, a model person divested of all human failings. He was utopia projected into the
individual, divested of human psychology. Conservatives, meanwhile, more interested in Practicalist agendas such as
strengthening industry and promoting moral restraint, tended to
prefer such lessons as "humility" or "patience." For conservative
authors, the scientist was more in line with Benjamin Rush's
"Republican machine," noble in his dedication to group purpose
and in his denial of passion.

Whatever their politics, leftist or right wing, authors of the
day found "science" a platform to launch their ideas. Dewey
himself, though by this time much interested in the question of
"the historical and social consequences, both good and bad, of
science" (and aware, as well, of certain problems inherent in the
idea of "disinterest"), remained a strong proponent of the notion
that "the future of democracy is allied with the spread of the
scientific attitude—[our] only assurance of the possibility of a
public opinion intelligent enough to meet present social problems" (1939, pp. 148–149).

The period from 1930 to 1950 was marked by effects to
popularize the image of the scientist as superman and science as
the path to utopia for the less well-educated reader too. One
such effort involved the rise of the science fiction as a pulp
magazine genre, achieving an enormous following in but a few
short years (its golden age was the 1930s and 1940s). Hugo
Gernsback, one of science fiction's founding fathers, admitted
that the true intent of this literature was pedagogic: to use
"astounding" or "wonder" stories as a means to teach science to
the uninitiated (Carter, 1977). Drenched in images of a spectacular yet troubled and conflict-ridden future (designed to stage
obvious conflicts of the present), this early sci-fi writing was

intended to champion the benefits humankind would receive thanks to "the scientific mind." Heroes in these stories were often everyday amateur non-scientists who used technical knowledge to rescue civilization and self. But the true protagonist was always "science" itself, the 'universal' power of progress, granted by reason, and given most of all to the powers of American democracy.

Sci-fi was aimed mainly at those who already had an interest in technical knowledge. Other Americans meanwhile, in whom such an interest might be germinated, a veritable torrent of books promising "science for everyman" appeared during these years, many of them written in response to the much declaimed "poor state of knowledge" among enlisted men during World War II. Most works in this category were simply bursting with the benefits to be derived from an individual grasp of what "science has to teach us." Books written expressly as friendly texts for the layman tended to sell best and attract the greatest amount of applause from experts. Among the most successful of all, for example, was a series on mathematics by H. G. and L. R. Lieber, especially *The Education of T. C. Mits* (*The Celebrated Man In The Street*, 1942). Written in an inventive form of mock free verse meant to visualize the very essence of simplicity and clarity, illustrated with whimsical, childlike drawings, this work wears its commitment to everyman comprehension on its sleeve:

> Oh, we know you do not like
> Mathematics,
> but
> we promise not to
> use it as an instrument of torture
> but to show
> what bearing it may have on . . . such things as:
> Democracy
> Freedom and License
> Pride and Prejudice
> Success
> Isolationism
> Preparedness
> Tradition
> Progress . . .

Human Nature
War
Self-reliance
Humility . . .
Anarchy
Loyalty
Abstract Art
and so on. (pp. 12–13)

The range of human phenomena to which the teachings of science are presumed to bear relevance is nearly unlimited, varying from the grandest scale to the most private and personal. Yet there was one other reason for being inducted into an understanding of these teachings:

No doubt someone will say:
"But the war-makers
DO use modern machinery . . .
Science is really to blame for the success of Hitler,
and therefore
it cannot possibly guide us to
the good life."
Now we hope to show here that this is not so—
that Science and Mathematics . . . have within them
a PHILOSOPHY which
can protect us from
the errors of our own loose thinking
And thus they can be
a vertible defense against
ALL evil. (pp. 48–49)

To such popularizers, therefore, a proper understanding of technical knowledge would save the individual from moral weakness, democracy from defeat, and science from unjust attack. By this time, in other words, in addition to everything else that was claimed for it, science was being promoted as a way to prevent, do battle with, and even *explain* evil (as the misunderstanding of "science" no less). True to its promise, this book, like nearly all others of its kind, ends with a series of ethical truisms. These wartime rules of conduct tended to be conservative, reminiscent of Roman stoicism: "Modesty and humility and self-reliance should characterize man's activity," "Clear thinking combined with careful observation are his most 'practical' weapon," and so on. As a source of higher morality, science would make better soldiers, men capable of winning all wars against all tyrants. All the ideals associated with the promotion of scientific learning in the late 18th and early 19th centuries—all its moral, intellectual, economic, national, and symbolic goods—are apparent here, in forms adapted to the historical circumstance of depression, world war, and general uncertainty.

Among educators, meanwhile, this discourse of promise and protection had its own particular spin. Here, for example, is a declaration of philosophy given by the National Science Teachers Association in 1942:

> In order to be functional science teaching must be directed toward better living. The scientist has demonstrated . . . that the scientific method, supplemented with careful study of contemporary living, can be applied to the solution of personal and social problems . . . *The spirit of science is the wish to know, the urge to seek, and the desire to comprehend the universe.* . . . Several reasons might be offered to explain mankind's lack of utilization of [the scientific method] for the solution of problems [to date]. [First among these is that] emotion interferes with clear thinking. The scientific method is rational rather than emotional. Undisciplined man is an emotional animal. (in Boenig, 1969, p. 12)

A full century after Horace Mann and nearly 50 years after William Torrey Harris, education is again proposed as a replacement of nature by culture—only this time the latter term is equated with its early nemesis, science. The highest form of civilized existence, what had once been associated with "culture" as an embodiment of classical wisdom, is now turned over to "scientific thinking." Indeed, the "solution of social problems" can only come about by the transformation of all Americans into scientists of a sort. Yet what does this transformation involve? As a utopian substance of mind, science is somehow about wishing (to know), about urges (to seek), and about desires (to comprehend); but it is also unemotional and rational. Such contradictory notions, so central to the positivist outlook, were bound to end up in a melange of simplification, idealism, concern, and authoritarian dismissal.

What, then, of the curriculum? As so often before, one might expect it to be heavy on the natural sciences. After all, if "the inclusive purpose of education is the democratic way of life," to be achieved through "self-realization, human relationships, economic efficiency, and civic responsibility" (National Science Teachers Association, 1992 p. 13), what could be more fundamental than those subjects that teach "clear thinking" over everything else? Yet one finds that the Progressive legacy continues to invert the relationship: Instead of learning biology or physics, one must learn health, safety consciousness, orientation to work, the promotion of intelligent consumership, maturation of interpersonal relationships, and so on. The *Cardinal Principles* were never abandoned, merely amended. Whatever heroics might be associated with "the scientific method," they were not to come from the teaching of science itself, since the goal of "better living" meant that education should be devoted to training almost every aspect of the person *except* the intellect (one sees, sadly, how Dewey's idea of reconnecting knowledge to its social labor was bludgeoned into "life adjustment"). It remains only to say, perhaps, that, the final role of schooling was to use science above all in ways that attempted to systematically make all students middle class.

This kind of thinking, it should be mentioned, had always been compelled by egalitarian motives. The traditional curriculum, based on the "liberal arts," was viewed as being skewed in favor of the minority: the intellectually gifted, the ambitious, or those of wealth and privilege who were already destined for elevated positions in society. A true democratic education, in contrast, had to be designed for the majority, the mass of "common students," with moderate or even inferior ability and interest, whose path through life could be helped far less by the exposure to abstract ideas than by direct training in "daily skills" for the solution of "real-life problems." The point of it all, later expanded to support the sweeping life-adjustment movement of the late 1940s and early 1950s, was to make the school not merely a social service but a special kind of social service, a kind of foster parentage. Indeed, the effort to create good family members, moral workers, and responsible consumers transformed the teacher into something like a surrogate mother or father. Certainly, teaching, as a social task, was aimed at promoting obedience to ethical norms. It could not have included the aim to inspire a questioning or critical attitude toward the "knowledge" being taught, or how the contents of this knowledge were determined.

The life-adjustment movement sought a state-sanctioned, paternalistic ordering of adolescent inclinations and behavior. It was not vicious, not terrified of a rising class of teenagers (as has sometimes been suggested). It was instead a well-meaning, literal-minded, anti-intellectualism that crashed on the rock of its assumption that democracy must always begin with the lowest common denominator—just as its historical opposite, traditional Academism, had previously lost power for its opposite presumption. That both these camps existed simultaneously amid the growing conservatism of the late 1930s and 1940s should not be thought surprising.

The work of Hutchins and his followers, and their success in restoring a core liberal arts curriculum at the college level, can be seen as an almost required counterpart to the continued reign of mass education for the high schools. College was to remain, more than ever, a filter for the higher individual. "One purpose of education is to draw out the elements of our common human nature," Hutchins wrote in *The Higher Learning in America* (1936, p. 166). "These elements are the same in any time or place. The notion of educating a man to live in any particular time or place, to adjust him to any particular environment, is therefore foreign to a true conception of education." This fixed and "common human nature" meant for Hutchins a refined substance, not a democratic one, more specifically defined by the "great books" of Western European urban culture, in which science presumably played little or no part.

Dewey called this idea an escape, but it was more a defense. In calling for a return to the previous century Hutchins was reacting against the disturbing diversities of "culture" (high and low) that he saw everywhere in evidence. His greatest fear, no doubt, was that the depressing state of affairs in the high schools would infiltrate the colleges. To oppose the tide of an ever-more differentiated curriculum, he defended a course of study that would define its very opposite. A core curriculum that returned to perennial wisdom would save higher education from any descent into social service. Hutchins may well have wanted to "save democracy" too—but this rescue would have to begin with the colleges. Only the colleges could revise contemporary society's inclination to promote the masses at the expense of superior individuals. Hutchins may have thought that his defense of elite education would save American democracy from totalitarianism. What Dewey more precisely understood, however, was that fascism depended less on mass culture and far more on the type of absolutist ideas Hutchins himself was propounding—ideas of eternal and unquestionable truth, which came to be embodied in despotic leaders.[6]

TRENDS IN SCIENCE EDUCATION

None of this can be separated from what happened in Academist science, in professional scientific education. During the 1920s, 1930s and into the 1940s an increasing number of scientists became involved in corporate and military work. This was aided in no small part by the use of scientists for the war effort during Woodrow Wilson's administration. In 1916, for example, Wilson endorsed the formation of the Naval Research Laboratory and the advisory National Research Council, stating that "preparedness" for the nation would have to be "solidly based on science" (Kevles, 1978, p. 102). More generally, the heavily mechanized nature of weaponry, the use of chemical warfare, and the consumption of large quantities of natural resources such as oil, coal, and iron guaranteed the deep involvement of industrial scientists and engineers in mobilization activities. During the 1920s, Herbert Hoover, himself a mining engineer by training, fostered many new connections between industry and the federal government in the name of scientific progress, especially though the Commerce and Interior departments. As a result of these and other efforts, the resurgence of "science for the national interest" led to a great expansion in federal and industry underwriting of the research programs in major universities—of the research university itself as *the* major social institution for the production of technical knowledge and the training of its experts. Such was the long-term effect of the Morrill Act, which had first politicized science on this type of concrete basis.

Between 1930 and 1950 the founding of many federally financed research labs, for example, the Argonne lab at the University of Chicago, the Jet Propulsion lab at Cal Tech, and Lincoln lab at MIT, was merely one example of a much larger and by this time well-established tradition that included many fields (but especially physics) in its overall grasp. Federal policy consciously followed the model set by corporations, which had set up as many as 375 research laboratories by 1917 and over 1600 by the early 1930s (Pursell, 1981). Such labs, in turn, had been built both in imitation of, and in competition with, the research universities themselves, and had proven over and over again the benefits of "pure inquiry." In founding its own university-affiliated laboratories, usually through military-related funds, the government always had pragmatic motives: Academist science in the name of nationalist ends was always the first and final word on the matter.

Teaching by means of research apprenticeship, by now well-established at the senior undergraduate and graduate levels, was the method used in all these federally financed labs. These labs might be seen to resemble in their half-independence the old scientific schools of the Ivy League colleges which in their day acted as professional training centers for scientists. The difference, however, was one of status and privilege: Even at their founding, many 20th-century scientists considered the new laboratory centers to be the historical climax of the successful university—founded on, built from, and now the physical embodiment of a national committment to technical progress. (Others, however, feared that work done in these labs would be coopted for military uses.) Indeed, federal involvement in research, begun on a large scale during World War I, created a climate that made a program like the Manhattan Project much less revolutionary than has often been claimed. The new labs, after all, were "Big Science" on a big new scale, and the ambitions involved were often claimed as commensurate. No longer was the goal merely to "purify" or to "coax" nature into revealing its secrets. Such manipulations were of a compromising order, appropriate to small teams or single investigators, awed by the scale of their study rather than by their methods and tools. Now, however, with huge technologies and entire bureaucracies at their disposal, scientists were on the side of power, able to "create" or "deform" nature into yielding new knowledge. The idea of "research" gained an even more sacred character than it had aquired in the past: Research work, that is, was the essence, the beginning, end, and middle of all science, no matter in what area. The search for fundamental truth was just as important as before; but now this search was impossible without its prefix.

Such changes in the character of knowledge, and the philosophy underlying them, did not bring with them any significant changes in the mechanics of teaching, as mentioned. For undergraduates, it still depended above all on styles of ingrained imitation, with strong reliance on the close reading and memorization of textbook material, cookbook lab work, and recitational exams. Students in fields such as chemistry, physics, and geology might be enlisted to help work on a project sponsored by a specific corporation (or by a professor engaged in related consulting work), such as Dupont or General Electric. But this usually involved performing menial tasks (simple titrations or other chemical analyses, measuring specific gravity of liquids or minerals, and the like), according to formulaic procedures. At base, Academism in scientific education for scientists was never really altered in any significant way, not by Progressive and neo-Progressive reforms, not by the "science for democracy" movement among educators, not by scientists themselves. The ascent of research to ecclesiastical stature only helped deepen the authority-driven relations within classroom and laboratory. The apprenticeship hierarchy remained entirely in place, with original work being confined to graduate study— the final gateway to professional standing. Scientists seemed to feel that the existing system worked well, that it produced competent, even brilliant technical workers and thinkers— indeed, didn't the adulation being given science in various parts of society confirm this?

The most introductory courses in colleges and universities were full of such Progressive sentiment. Such courses, meant to spark an interest in science among undecided freshmen, were taught via the old potted histories of great minds and their discoveries, along with rudimentary laboratory skills, tinctured through with the same type of proselytizing rhetoric discussed above with regard to the science-for-the-everyman movement. Some flavor of this rhetoric, for example, can be seen in the following excerpt, taken from a much-used text of the time:

> Scientific knowledge has changed man's systems of thought and the manner in which he views the universe. In many respects it has altered his whole attitude toward life and the meaning of his existence . . . Science has also profoundly influenced man's conception of dependable truth and the method of arriving at it. It has introduced a new method into rational processes and has laid down specific rules which man must follow if he is to reach valid conclusions. (Jean, Harrah, & Herman, 1934, p. iii)

Such prefaratory remarks, whatever their heroic intentions or mere reflexiveness, did not prevent such courses from being taught in frequent rote fashion, through the same old lecture-lab process. Scientists did not agitate for reform, perhaps in part, due to their psychological attachment to a rite-of-passage set of experiences. But it is more likely that they enjoyed, to a greater degree, the privileges of control over career-making, the high status of university professorships, and the often lucrative nature of those professorships with regard to both symbolic and monetary capital.

At the high school level and below, meanwhile, where students with potential were first identified (and tracked), other reasons must be invoked to explain the lack of change in science teaching. These have been discussed in a broader context, for all of teaching, by Cuban (1984) for the entire period 1890 to 1980. His arguments, in fact, based on extensive review of teaching practices, seem at the very least persuasive and strongly reinforce Dewey's own perception of how school structure can defeat reform. For Cuban, the details of instruction cannot be divorced from still larger structures, those related to authority and performance in the socioeconimic environment:

Those instructional practices that seek obedience, uniformity, and productivity through, for example, tests, grades, homework, and paying attention to the teacher prepare children for effective participation in a bureaucratic and corporate culture. Consistent with this argument is that certain teaching practices become functional to achieve those ends: Teaching the whole class encourages children to vie for the teacher's attention and encourages competitiveness; teacher questions reward those students who respond with the correct answer. (p. 9)

Cuban notes how institutional controls are made more rigid by the organization of school experience itself: for example, by such things as age-graded classes, the division of the school day into time-clock periods, the physical layout of the classroom, the dictated nature of extracurricular activities, the emphasis on assessment and teacher evaluation, and so on. He also mentions peer pressure as having its own educative power within the teaching profession, one that works toward stasis. New teachers, whatever their training or ambition, are often caught up in the existing social network of a particular school, and urged to learn by observing their elders; if they do not adjust to "the way things are done," they are punished by means of their job assignments, ostracized by other teachers and administrators, or even fired. Cuban also notes that whenever reforms have been introduced into a school system, the changes have been implemented in an unsystematic way, due to inadequate resources, political parochialism, or poor understanding of the rationale for the reforms vitiate their effectiveness.

What Cuban does not discuss, and what may be at least as determining as everything else, is the simple fact that Progressive reforms were not often as child-centered as they claimed or hoped to be. On the contrary, as we have seen, they very often acted to deepen or expand still further the fundamental structures already in place, structures that now acquired privilege by being wetted all over with a film of the "scientific."

This is not to say that change was altogether absent, or that the reforms that did take place were entirely trivial or temporary. Even by 1915, the entire system of 19th-century instruction based purely on rote learning was largely eliminated.[7] Desks bolted to the floor were mostly gone by the 1930s. In the sciences, individual lab work became relatively commonplace at about the same time. Field trips, personal essay writing, even group learning and independent study had come to exist in some of the nation's high schools by 1940. At the elementary level, moreover, there grew up an actual tradition of founding local experimental schools, especially in the 1910s and 1920s, each of which tried to design a curriculum that would (finally) place the child first, and give freedom back to the teacher (see, e.g., Elias, 1973). And above everything else, the Progressive vocabulary steeped in "growth" and "learning as living" could be found on the lips of most teachers by the 1930s.

The problem was that these changes were largely epiphenomenal. As Cuban (1984, p. 32) points out, they did not prevent "the connective tissue of instruction—classroom architecture, class size, report cards, rules, evaluation process, and supervision" from remaining fixed. Teacher talk, teacher-dictated use of time, and teacher-delivered instruction, aimed at the whole class (or repeated to individual groups), dominated the scene no less than before. Moreover, the freedom granted teachers themselves to implement change was often not forthcoming, and always reliant on higher authority. As Dewey himself noted, as late as 1952, "The success in the teacher–student relationship [*sic*] remains limited; it is largely atmospheric. The older gross manifestations of . . . education by fear and repression—physical, social, and intellectual—have, generally speaking, been eliminated. . . .[Yet] the fundamental authoritarianism of the old education persists in various modified forms" (in Dworkin, 1959, p. 129).

The main agent of change, then, was the curriculum. And what did it alter, in the end? Not the methods of instruction, nor the relative experience of teacher and student. What it reformed, as the instrument of Progressive intent, was the attempted reach of the school into nearly every aspect of the life of its pupils. For it was this reach, with the teacher as its hand, that was given new power over a vastly increased population. It was a reach too that neither extended nor retracted for decades thereafter. For, to any significant degree except loss of academic course work, the curriculum itself stopped changing after the early 1920s, and remained fixed in its life-adjustment content until well into the 1950s. The old Committee of Ten curriculum was at its peak around 1910, when nearly 50% of public high school students were enrolled in Latin, and upwards of 80% were required to take mathematics and science. By 1949, these figures had fallen to about 8% and 55% (if "general science" is included). In modified form, the committee's curriculum contined on for upper-track students, which comprised some 25% or less of the total. For all others, as Hofstadter contended, the demands of the life-adjustment advocates had become "practically insatiable" (1963, pp. 337–358). This was therefore posed to spark a rebellion in the opposite direction with the coming of cold war fears.

But in the 40 years before 1957, the year Sputnik was launched, life adjustment held firm as the curriculum standard, a remarkable length of time given the changes that took place in American society. Its success was to provide the outward forms and rhetoric of "democracy in the classroom," while continuing the old Academist teaching without pause. During this period, in fact, an important chasm came to divide pedagogic theory from practice: Debates that raged over goals and philosophy—for example, those surrounding George Counts's 1932 book *Dare the School Build a New Social Order?*—barely had any influence at all beyond the podium. Counts' book, which raised a sensation within the pedagogic community, began with a castigating overview:

> That the existing school is leading the way to a better social order is a thesis which few informed persons would care to defend. Except as it is forced to fight for its own life during times of depression, its course is too serene and untroubled. Only in the rarest of instances does it wage war on behalf of principle or ideal. Almost everywhere it is in the grip of conservative forces and is serving the cause of perpetuating ideas and institutions suited to an age that is gone. (1932, p. 4)

Progressive education, Counts said, once properly understood and best implemented, had the capability to begin a new and better era. In a much-quoted phrase, aimed squarely at neo-classicists such as Hutchins and Flexner, he stated "If schools are to be really effective, they must become centers for the building, and not merely for the contemplation, of our civilization" (p.

13). As yet, however, Progressive education did not approach this ideal; its weakness lay in the fact that "it has elaborated no theory of social welfare, unless it be that of anarchy or extreme individualism" (p. 5).

In Deweyan tones, Counts wrote as if the original vision behind Progressivism were a new thing, untried and gleaming with promise. In fact, his diagnosis, which sparked several years of intense discussion, was almost the reverse of the truth: The "new education" was not at all about "anarchy or extreme individualism" but about the very opposite, the designing of a mass individual for the building of a civilization where everyone, in essence, knew and obeyed their place. Based on child-study and on "scientific curriculum making," reformers had established for decades a philosophy that conceived the subject of learning as a generic entity, waiting to be guided, led, or shaped. Counts's misreading of the larger circumstance of schooling reveals, in part, how academic such discussions had become, how pedagogy was now itself a self-contained profession, a matter of "research" at the highest levels, unable to percolate downward.

In the end, the great irony of the Progressive movement in education was its very success. It had succeeded in completing the project of a mass education, and it had done this under the auspices of an ever-more disseminated yet no less massive system of authority dispensation. The school-as-factory image is one that came to be rejected by increasing numbers of educators. It was harsh, it was exploitive, it was uninterested in the needs and hopes of each individual. But most of all, it was accurate, and herein lay the force of the present. From now on, though Dewey's rhetoric would have its frequent recurrence, the more guiding word in practice was that of "skills," the true indicator that the student had become a kind of laborer for a system of social training. The continued success of Progressivism was the establishment of a clear historical fact: that however much learning might remain an individual affair in its essence and experience, education was now a full-scale institution, with everything that came with it.

The Postwar Era:
The Return of Academism
and the Sputnik Revolution

Late in 1958 an inauguration ceremony was held for the new president of the University of Washington, Charles E. Odegaard. In attendance were some of the most eminent scholars and college administrators of the postwar era, men from both the sciences and the humanities who had been invited to speak on one of the most pressing topics under their purview, "The Rededication of the University to Its Function." Among these scholars was the renowned Johns Hopkins University philosopher George Boas, who began his address with the following provocation:

> Professors of the humanities are in a great tizzy nowadays. They see the scientists getting all the money, research grants, big fellowships, government contracts, and of course large salaries. They are told that students are drifing away from them to enter the natural and the social sciences. They seem to have the feeling that their work goes by the board and that they are becoming vestigial organs on the body academic. As far as the facts go, there is a good bit of truth in this. Whether it is something to be deplored is another question. But if humanistic studies are becoming obsolete, one might at least ask why. (in *Man and Learning in Modern Society*, 1959, p. 126)

Boas's own answer, delivered in the very next sentence, was no less barbed: "If humanists think that modern society has rejected them . . . they might ask whether it is not they who have rejected modern society."

As far as the facts went, however, Boas expressed not so much the truth of the matter as what many academics felt. Indeed, the sciences had made great strides in the arena of privilege. The war, in fact, had done much to ensure progress in this area. Even the old Ivy League colleges were now fully modernized, on a par with the best research universities. Outside, it was the "jet age," the "atomic age," the "age of automation," the era in which technical knowledge was both the source of greatest threat and the holder of maximum promise. Inside, too, it therefore seemed to be the "age of science" in higher learning, a time when the arts were becoming a mere elective in a present beyond their control.

The perception was true in some sense, but false in another, because of its very premise. The rise of science, now deeply integrated into the "liberal arts," was linked to the rise of Academism generally, in all its precincts. A full 20 years before the launching of Sputnik in 1957, strong arguments had come forward against the (now traditional) Progressive view of secondary and higher education—arguments calling for a required core curriculum of "fundamental" subjects, for teaching common values, and above all, for the "return of wisdom." Led by the energetic and vitrioic Hutchins, the "Essentialists" had important influences on the curriculum, including the implementation of "great ideas" survey courses, honors programs, and other changes. But such things did not at all satisfy Hutchins and his followers, who wanted colleges swept clean of all professional training. Their voice grew still louder and more insistent by the late 1930s, as the country and its colleges swung to the right. Throughout the 1940s and 1950s, even into the early 1960s, Hutchins continued to sound his message without quarter, but with an ever-increasing note of urgency and sarcasm. In 1953, he published *The University of Utopia*,[1] a diagnosis and cure regarding the "peculiar dangers" facing education in the United States. These he claimed were several:

> Industrialization seems to charm people into thinking that the prime aim of life and hence of education is the development of industrial power. Specialization has dire effects upon the effort to build up a community and particularly a community of the learned. Philosophical diversity raises the question whether a community is possi-

ble. Social and political conformity, on the other hand, suggests that
the kind of community we seem to be headed for is one that we shall
not like when we get it. (p. 1)

As before, the solution to all these intransigent difficulties was
simple: "What we are looking for is wisdom, and it does not seem
sensible to say that the insights and understanding offered us by
the greatest creations of the human mind cannot help us in our
search" (p. 16). To the question "What were these creations,
exactly?", Hutchins had an equally simple answer: "It would be a
bold man who would say that Newton had taught the West more
than Shakespeare" (p. 16). Indeed, Hutchins spends no small
portion of his book relegating the sciences to "the material
conditions of existence," thus claiming their irrelevance to moral, spiritual, or other intangible yet essential aspects of "wisdom."

Hutchins's overall program, like that of other similarly-
minded educators, smacked of 19th-century classicism in its
more raw form. Yet it nonetheless included the sciences as an
integral part of the "liberal arts" (though science had helped
"trivialize other fields of learning," it was still, in its purer
research form, "perhaps the greatest accomplishment of modern
times," thus some knowledge of it was essential to the possibility
of wisdom; see pp. 15–16). Hutchins was thus by no means
alienated from the framework of higher learning as it had come
to exist. Ironically, perhaps, his call for increased "rigor" and
"intellectual virtue," his insistence that "patriotism, moral fervor, and intellectual capacity [provide] the ability to meet any
new situation with intelligence and decision" (p. 21), tended to
aid and even enhance demands in the post-war era for a more
difficult, challenging type of training, in the sciences especially.
Greater knowledge for a wiser nation went hand-in-hand as a
slogan sensibility with stronger research science for greater international power. Indeed, the "two cultures"—the humanities
and the sciences—may well have fought over their respective
turf and relative merits at the academic roundtable during this
period. But with regard to the most basic aspects of curriculum
philosophy, they were in complete agreement about the need for
a purist, Academist ideal.[2]

Indeed, their very conflict in other areas aided such agreement in the end. The call for "wisdom" gained a new imperative in the 1950s, for it was now predicated on the unexpected and disturbing uncertainties injected into Western society as a result of scientific work in military hands, the atomic bomb most of all. There was no surprise, therefore, that one of the participants at the Odegaard inaugural who spoke most fervently (and characteristically) of the deeper vocation of the "arts" was himself a well-known physicist, Polykarp Kusch of Columbia University, who stated:

> Whether science in the present world will elevate the human spirit and . . . ennoble man, or whether [its products] will lead to a destruction of life . . . and to the complete degradation of the human spirit, is the crucial issue of the age. The issue will be resolved by men; I myself will be much more optimistic of the future if the men who resolve the issue are men of wisdom; not scientists as such . . . but men who can deal sagaciously with knowledge. Therein lies, I think, the destiny of the university . . . to produce men of wisdom. (in Hutchins, 1953, p. 69)

Wisdom, in short, now meant survival, and it was not the exclusive domain of science. It came more from "a knowledge of man, of his history, of his aspirations, of the good life . . . of the nature of human society . . . [and of] a sense of beauty" (p. 66). Above all, as embodied in the university, it came from a dedication "to intellectual values." The Academist ideal had indeed returned, and with new gravity. Somehow, once again, the world would be saved by the giving of "culture" to those who would be leaders, scientists, or otherwise. Somehow literature and philosophy were, in their essence, on the side of democracy.

Contexts and Influences

The forces driving a renewal of this ideal need to be placed within a context of influences. These include the pressures exerted by the GI Bill, by the new scale of federal support in science, and of course, by the cold war. Taken as a whole, the argument for curriculum change in this period was one based less on positive feelings than on negative ones, especially those arising from one or another type of fear, whether related to international strife or the mass popularization of higher learning.

One should note, first of all, that the high schools and colleges had become increasingly divided in terms of their curriculum philosophy. While life-adjustment curricula continued nearly unabated in the former until the mid-1950s, many universities had already reinstated a standard liberal arts course of study. This difference had everything to do with how students in each category were viewed. High school was now populated by "teenagers," who were the objects of general concern. Worries about delinquency, about fitting them into the social order (by means of better parenting, vocational training, providing jobs), behavioral norms, and so forth, remained paramount. College students, on the other hand, were viewed as "the leaders of tomorrow." As such, they were destined for specialization, for expertise. But the war, along with fascism and Stalinism generally, had cast doubt upon the value of expertise, revealing that experts could be immoral and mercenary if not guided by "higher ideals." In the Western democracies, the expert required a broad yet structured foundation, a "general education in a free society." He or she required a kind of training that would not lose sight of a greater mission, a training that would help shape his or her sensibility to use the power of knowledge for the good of all.

One might say that the teenager and the expert were two sides of a single coin. Both required control, which education could provide. It seems interesting that the more powerful of the two was to be checked by "intellectual virtue," of the standard liberal arts kind, while the other could simply be contained by a condescending custodial system. This dichotomy reveals, however, the strength and style of belief involved. The deeper lessons to be gained from study in these arts, from their elevated culture, was to include various forms of restraint—a curbing of runaway individualism, a channeling of ambition into higher causes, a faith in "civilization [as] the deliberate pursuit of a common ideal" (Hutchins, 1953, p. 52). In short, what the liberal arts would now provide, for both survival and the betterment of society, included everything that had been reserved for "science" only a decade or so earlier. The shift of the moral high ground from "the scientific attitude" to "the liberally educated individual" marked an enormous loss of faith in one area, and a de facto rise (of less magnitude) in another. Both within academia and without, the use of the atomic bomb, the initiation of the arms race, and nuclear terror helped ensure that "science," as a grandly reified entity, would never again be divorced from its destructive possibilities.

This does not mean that the public and academic image of science was wholly tainted, that science was now viewed as a matter of "evil wizards" (Allen, 1992). The notion that the prestige associated with technical knowledge fell dismally during this period is untrue, even absurd. The 1950s and early 1960s were a time of vast popularity for scientific effort and accomplishment, a time when enrollments in technical disciplines increased enormously, when chemistry sets and microscopes sold by the millions, when "the wonders of nature" and "science for tomorrow" appeared as common motifs on television, in magazines, in advertisements of all sorts (even encyclopedias altered their format and subject matter to focus more fully on scientific subjects during this period). Certainly fear and suspicion existed, but in highly complex and shifting ways. Distrust was aimed sometimes at "scientists," often at "amoral" or "hired" science, or at "science in the wrong hands," frequently at "Russian science," or even at the bad "side effects" or "undue reverence" that scientific knowledge engendered. For every classicist such as Hutchins, who blamed technical knowledge for its uses, there were a dozen others who spoke instead of the abuse of such knowledge (for nonscientific ends), of the need for "good science" to counteract "bad science." In short, attributions of responsibility were extremely variable and tended to reflect a speaker's political interest more than any general attitude among the American public. Overriding all else, too, was the bomb, which acted as a kind of totemic magnet, drawing criticism that might otherwise have been directed elsewhere.

The decline in faith, though enormous, was restricted. Science had been divested of its spiritual content, of a necessary moral superiority. And this was of no small historical significance. Never again would "the scientific mind" be the golden braid to a better world. But the true loss of innocence was still yet to come. Few, if any, critics had yet to attack the idea of "objectivity" itself, to place scientific work and thought into a contingent, social context. Technical knowledge was still viewed as a bringer of wonders, as a bearer of human progress and reason. But now, commensurate with new uncertainties, it was often envisioned as having amoral capabilities, as a weapon or force to be directed against enemies. Of course, it had been viewed as a weapon before, during the World War I especially. But at this earlier point in history, science was, if anything, a kind of heroic agent of necessary death, a tool for destroying the threats to democracy and freedom. More Americans saw it as being on their side above all—they believed the progress of their nation had given them a claim on science, as the "field at their door." Thus there was never any conflict between science as an instrument of war and as a moral/spiritual champion for American democracy. But now, it was clear, any nation could make use of technical knowledge. Now, science, for many, was the origin of internal and external threat, and therefore the means to overcome it as well.

Against this background, then, several influences worked to encourage conservative changes in the curriculum. The war had greatly accelerated the scale of professional research science. It had also, by means of military industrialism, provided the government with its main engine for stimulating the economy to create a new era of (selective) prosperity. The combination of these two factors led to what Clark Kerr, president of the University of California, later called the transformation of the land grant university into the "federal grant university" (Kerr, 1963). In federal and industrial labs, as well as within academia itself, employment opportunities burgeoned, and the call went out for more students in the sciences, especially those with professional ambitions. At the same time, the notion that a more demanding curriculum in science would help pupils better understand the brave new world they inhabited became widespread. The result, again, could have been predicated by anyone familiar with educational history. Vocational and general science courses in college—the legacy of life-adjustment—were found to be inadequate. By the mid-1950s they had been largely replaced with a more strictly academic curriculum, in which each field offered its own introductory courses. Some schools, prior to Sputnik, added the demand the students in specialized areas, such as engineering, take 10 or 15% of their course work in other, nonrelated fields. This was in compliance with the demand to provide a more general, liberal education to future industrial leaders. It represented an academic ideal no less than a practical one.

A second influence in this direction, also related to the war, was the huge increase in college enrollments brought on by the GI Bill in the late 1940s. This federal effort marked the final stage in the spread of mass education in America. From now on, higher learning would continuously broaden its mandate to educate, first by absorbing ambitious war veterans, second by making tertiary education the expected goal of most middle-class students, and third by extending its reach even into the lower classes. But in the beginning of this process conservatives hoped that higher education's outreach would be a short-term phenomenon, that things would subside back into their old selective patterns with a few years. This did not happen. On the contrary, from a total of about 750,000 students in 1940, enrollments swelled to 2.3 million by 1947, and then simply kept on growing, to 2.7 million in 1950, and then to nearly 4 million by 1960 (Cremin, 1988, pp. 251–252). A great deal of discussion took place at the upper levels of academe concerning this growth, even during the war years. Whatever their previous commitment to "defending democracy through education," many conservative administrators and educators now perceived in this sudden influx of students a possible diluting force to the quality of higher education. When incoming veterans did much better academically than anyone had expected (especially the existing professoriat), conservatives used this information to argue that standards were too low—after all, how else could one explain seeming achievement by an obvious nonelite? Later, these perceptions would be extended to the high schools as well, not so much because enrollments there increased proportionally, but because this level of schooling ceased altogether being a filter that separated the masses from the elite destined for college, instead becoming a site at which multiple future tracks were determined. Sooner or later, it was inevitable that college-type curricula and expectations would be forced downward, both in the name of higher standards and national strength. And this is exactly what happened, following the Sputnik "crisis." Indeed, it was proposed in no uncertain terms by Clark Kerr at the Odegaard conference of 1958:

> We are going to have to undertake more intensive education of all our youth, not only at the university level, but I would say particularly in the high schools of our nation. I am quite convinced that we can push back a good deal of the work that now must be given in the freshman and sophomore years of college to the high school level, thus making it possible to give expanded and enriched training in the four undergraduate years. (in *Man and Learning in Modern Society*, 1959, p. 13)

Because of the challenges and difficulties facing American in the post-war era, Kerr said, "It is quite clear that we are going to have to do more in the field of education than we have ever had [to] before—more teaching, more research, more service" (p. 13). And what this meant, in particular, was "the necessity to turn our universities, particularly our great universities, from an over-emphasis on the vocational side of education and to make education more truly intellectual" (p. 14). This necessity existed, finally, because the "complexity" of the modern era created a "consequent need to train a new elite to run the world," an elite based not on "wealth and family, as in the past, but rather it will be an elite of talent" (p. 14). The belief that America was now meant to "run the world" reflected back the ancient demand of the imperial state to "educate the masters." Stricter standards and a type of training "more truly intellectual," moreover, would provide a means not merely to educate them, but to select them from an early stage onward.

Finally, the reality of the cold war exerted a powerful influence on curriculum content, helping force things in an Academist direction. Cold War rhetoric affected policy well into the 1960s, acting as an impetus to encourage more students to enter the sciences and engineering and to do so in more demanding courses. The intent here, in essence, was much less the

Progressive goal of teaching science to everyone for the purposes of better living than to get a larger pool of potential technical talent from which to create a more expanded, effective elite, one more capable of winning the race for international power, prestige, and security vis à vis the Soviet Union. This was where, again, certain purely nationalist aims, dependent on the emotion of threat, merged well with the Essentialist classicism of those such as Hutchins, whose emphasis on the need for "patriotism, moral fervor, and intellectual capacity" (1953, p. 21) were paramount. In a curious twist of history, cold war rhetoric regarding science merged with Hutchins' own sense that Americans as a whole had gone intellectually soft and were "culturally underprivileged" (1964, p. viii). Note, for example, the words of another participant at the Odegaard conference, Erwin D. Canham, then editor of the *Christian Science Monitor:*

> In terms of our national society itself, we certainly need to be reminded of the grim urgency of the challenge we face. It is survival. . . . We failed, it seems to me, [in underestimating] the capabilities— in science, technology, and morale—of the Soviet Union. We seem not to have kept up in the armaments race. . . . Deeper still was our failure to stiffen the will and strengthen the values of our own people. In the words of Senator Fulbright, these years of the 1950's have been marked for us by a weakness for the easy way. (in *Man and Learning in Modern Society*, 1959, pp. 40–41)

What this meant, then, was that "the people of the United States must understand their role in the world . . . and live up to it. Such understanding is manifestly a task of education" (p. 45). The final message, therefore, was always clear, always sooner or later a "war for the minds of men." Most of this would be waged over science, but not all. There were also calls for more experts in foreign languages (especially Russian, Chinese, Arabic, Spanish, and Hindi), in political science, in economics, in industrial design, and (in the wake of Korea and fears about "brainwashing") psychology. The "war" was to be fought, that is, with a full complement of the expertise needed for global influence.

The overall effect of these influences was that Practicalist demands were continually made for an increase in the scale of Academist science. For the research universities the 1950s and

1960s were a time of rapid growth. Direct support from both federal and corporate sources helped expand not only "basic" programs but new, larger facilities for "applied" research as well. Physics, chemistry, and electronics were especially favored, since they were viewed as far more likely to make important contributions to industrial and military strength. Yet civil and mechanical engineering also received major support, for example, thanks to the huge federal interstate highway construction project, and research work in applied organic chemistry gained large corporate support following the "age of plastics," launched during the late 1940s. As a direct result, as Cremin (1988) notes, "The university moved ever more to the heart of the economy and the polity, advising officials, credentialing and canonizing experts, and standardizing vast domains of intellectual life, from professional curricula to scientific classifications to information systems" (p. 561).

With the new influx of federal dollars, colleges all over the country launched a major phase of new building: Science and engineering "quads," new and updated laboratories, research centers, and libraries went up everywhere, enveloping the old liberal arts architectural core with a proliferating series of facilities almost entirely devoted to the sciences. Stylistically, the new buildings with their Bauhaus steel, cement, and glass design presented a jarring contrast to the older neogothic or Richardsonian forms, so reminiscent of an age when the humanities were still worshipped under the sign of "mental faculties." Now, however, the university came to be the very symbol of "big science" and everything it was to mean in late 20th-century society. It stood as the very opposite of the Deweyan ideal—not a community (embryonic or otherwise) of likeminded investigators, but a corporation whose product was "research" and whose operation took place according to corporate norms, with a hierarchy, a bureaucracy, competition, investment, attempts to increase market share, and desire for profit (both financial and symbolic).

This all had an effect on the basic structure of teaching. Introductory science classes became larger, more anonymous, and less interactive. Teacher–student contact was neglible.

Courses were taught by means of impersonal lectures, with lab work directed by graduate students. Successive years within a particular field were oriented toward "weeding out" nonmajors and encouraging those with higher ability, motivation, and professional ambitions. The tenure system, meanwhile, was transformed into a competitive bid process based on productivity: "output" in terms of research publications and ability to secure grant money. This change, moreover, was coupled with the university's own system of charging "overhead" against all individual research grants in order to expand and improve its facilities and thereby upgrade its own status to attract a more elite faculty and student body. Such changes, in short, went a long way to ensure the ultimate power and privilege granted to research over all other types of work, and certainly over teaching, both as a responsibility and as a possible career for field majors. Professors known to be "good" teachers were often saddled with greater course loads; in some schools, a conscious policy grew up to hire several such professors in each field so that those engaged in "important" research could be protected from having to "waste time" teaching undergraduates. A new kind of guild structure came into being.

The logic that led to the denigration of teaching and the valorization of research was directly transferred to the experience of students. Competition here was often no less severe or merciless. Viewed as the gateway to a successful scientific career, graduate school admission became the focus of an intense prestige/reward system based on grades, test scores, professional connections, and the like. Beyond this lay the highly restricted access to various "star" professors, whose work was shrouded in the glamor of "pioneering research" and whose fame, on a more practical level, virtually guaranteed the chosen student an advantage with regard to future placement. Finally, there was the growing battle to win research posts—whether in government, industry, or the universities themselves. During times of relative prosperity, this was less of a problem, as the number of positions tended to exceed the number of graduates. But during periods of recession, such as in the early 1960s, it meant fierce competition and meritocracy. More than ever before, higher learning became directly dependent upon changing economic conditions—an inevitable result of the university's transformation into a research corporation.

All these changes were reflected in the new official language surrounding science. In truth, much of the relevant discourse surrounding the notion of "technical manpower" was of the crudest type. Phrases such as "more mental muscle," "brains for freedom," and "more scientists means more science" were common refrains associated with a view that saw science mainly or entirely in terms of national service. And professional educators, again, were no exception. "Even in 1956, which was a pre-Sputnik year," wrote one, "it had become apparent that the national output of new scientists, technologists, and technicians . . . was inadequate." As a result, "the charge upon the schools [must be] clearly to provide science programs *for all students* that foster the development of scientific literacy. . . . [For] today we do not hear so much about keeping ahead of our world competitors, but we hear a lot more about catching up with them" (Fitzpatrick, 1960, pp. vii–viii).

Whatever the context, scientists were viewed as intellectual soldiers. To the degree that they would build and power the new war machine, they would also increase their ranks for the battle to "push back" the "frontiers" of knowledge. Materially speaking, the connection was obvious: bigger departments, bigger laboratories and research teams, more sophisticated equipment, higher salaries, increased competition—all these were a direct effect of growing federal and corporate involvement in research and development for military-industrial purposes. The entire organization of scientific practice was pushed in this direction by such involvement, which imposed the demand that the arms race and the profit race would parallel a "brains race." Some writers bemoaned or criticized this state of affairs in no uncertain terms—not for its trivializing of the humanities but for its attitude toward science. Regarding the entire Practicalist agenda of the day, for example, Paul Goodman wrote:

> The nadir of this kind of pitch for science has been, I suppose, the calamitous sentence in the late President Kennedy's message on

Education of 1963: "Vast areas of the unknown are daily being explored for economic, military, medical, and other reasons." (The "other reasons" include those of Galileo and Darwin.) Neither the scientific conduct of life, nor any conduct, is thought of as a purpose of education, or thought of at all. (1964, pp. 114–115)

Goodman's criticisms, however apt and pointed, nonetheless ended up in support of the old discourse of "objectivity" and "striving after truth" that remained firm among scientists, who otherwise profited nicely from the "brains for freedom" movement. At times, the rhetoric of a purist science was also enhanced by other mythic and often theatrical euphemisms that might pose science as "unlocking mystries" or "solving riddles" or "unveiling answers" or teaching a "higher conduct of life." New Images were added or stressed too, especially those that focused on one type or another of strength. More than ever before, technical knowledge now gave "power over the material world"; it "placed the forces of nature into human hands." At the Odegaard conference, the physicist Polykarp Kush expressed it thus: "I do not think that the most imaginative physicist of a century ago would have dreamed of the physical phenomena that are the basis of the present science of electronics. . . . The really important changes that occur are those that grow from wholly new knowledge, wholly unsuspected power" (in *Man and Learning in Modern Society,* 1959, p. 52).

The effects of war, hot and cold, upon the curriculum and upon science were penetrative. Increase and expansion were everywhere at hand, in the numbers of new students, in the scale of possible endeavor, in the tenor of competition. What occurred in education, generally, was a multileveled response to this situation.

THE CONANT INITIATIVE

James Bryant Conant, president of Harvard, was the high priest of Academist policy from the late 1940s until the early 1960s. Though Hutchins was a well-known figure on the academic scene, and was not without his influence, his voice by now was a wholly familiar (and somewhat monotonic) one, a known and recognized carryover from an earlier time. Conant, by contrast, was very much involved directly in the new era. His attitude and tone had none of the snobbish fatalism of the Essentialists. He was much more of an activist, an advocate for what he considered to be positive change in a forward direction. In 1945 he edited an influential volume entitled *General Education in a Free Society,* in which a group of Harvard professors addressed the subject of reforming both college and secondary training. The book was one of many published during and just after the war as a result of continuing debate in the colleges over the need to do away with Progressive reforms. The discussion in the Harvard volume is notable for its high-toned, reasoned arguments against the "enormous variety in method and aim" at the college level, said to have robbed liberal education of "any clear, coherent meaning," and for its questioning of the diversification of the high school curriculum: "How far can such paternalism go without sapping the final responsibility of the individual?" (p. 100).

The report addressed one of the most pressing issues of the day: how to strike "the right relationship between specialistic training on the one hand, aiming at any one of a thousand different destinies, and education in a common heritage and toward a common citizenship on the other" (p. 5). The "supreme need of American education" was for "a unifying purpose and idea" (p. 43). And to this, the Harvard group offered a "theory of general education," by which students would be instilled with "tolerance not from absence of standards but through possession of them" (p. 41). A large part of this theory revolved around the acquisition of "traits of mind," an updating of the old system of mental faculties in terms of such categories as: effective thinking, communication, the making of relevant judgments, and discrimination among values (pp. 64–73). These traits were said to be "necessary for anything like a complete life" (p. 79), thus implying (as all such documents eventually do) that scholars and professors—those who have seen the truth of such traits and their importance and who pursue the "life of the mind"—live more "complete" lives than do others.

Furthermore, despite claims made to contrary, such traits end up greatly favoring the humanities. Indeed, the report is filled with classical Academist calls upon the past: "It is certain that, like religious education, education in the great books is essentially an introduction of students to their heritage," and that this heritage, including democracy and its implications, involves "the long-standing impulse of education toward shaping students to a received ideal" (p. 44). Science, however important, is less a creator than itself a recipient of this ideal. "Specialism is the means for advancement in our mobile social structure; yet we must envisage the fact that a society controlled wholly by specialists is not a wisely ordered society. . . . The very prevalence and power of the demand for special training makes doubly clear the need for a concurrent, balancing force in general education" (p. 53). The sciences are thus stripped of any moral or social lessons. Instead, they are described as "a distinct type of intellectual enterprise, involving highly restricted aspects of reality," and (in Thomas Huxley's by-then platitudinous phrase), as "an expert and highly organized common sense" (pp. 150–151). The sciences "describe, analyze, and explain"; the humanities, however, "appraise, judge, and criticize" (p. 59). The realm of the former is truth, in naked, measurable aspect; the latter deals with "value," with "ideals of action, passion, and thought" (p. 60).

Besides this removal from the plains of wisdom, the Harvard document spoke of science education in a particular fashion. Much of its discourse seems defensive in character, in obvious reaction against life-adjustment notions: "The thought that an understanding of science might be conveyed as well or better without direct observation, experiment, and mathematical reasoning involves a fundamental misapprehension of the nature of science" (p. 153). And yet, at the same time, true to the Academist canon, only three fields are ever mentioned: physics, chemistry, and biology (always in this order), the "hard" sciences, which therefore stand in for the whole of technical knowledge. There is mention that "the facts of science must be learned in another context, cultural, historical, and philosophical" (p. 156), in other words that science must be humanized toward the goal of "wise understanding." Yet when it comes to actual methods and content, only a strict, traditional, academic-type course is advised, for example "study of the work of great biologists—Pasteur, Mendel, Darwin, and Harvey, for example—and

. . . through individual projects involving laboratory or field work which run parallel with the work of the classroom" (p. 159).

The intent thus seems less the teaching of science per se than teaching a certain image of the scientist as a removed and committed "investigator" proceeding in a cultural vacuum from experiment to hypothesis to discovery. This is the Academist image of the scientist, of course, by now a time-worn convention. But it was all the more false (and therefore all the more needed) in the new era of wartime science, grant science, and competitive science. Indeed, the gap between the reality of this imaginary lone "investigator" and the real-life research team was not merely large, it was iconic, since the former was always a means by which the realities surrounding and dictating the latter could be hidden from view.

Finally, the solution offered to reconciling specialization and generalism was not only less than radical or especially innovative; it was already tradition at the college level. Recommended for high schools above all, it amounted to establishing a "core of general education" that would occupy about 50% of a student's curriculum and would include a total of eight units, consisting mainly of survey-type courses, of which there would be three in English, three combined in science and mathematics, and two in social studies. This closely resembled the curriculum of the "junior college" in the 1920s and 1930s. More, it was a slightly watered down version of the old Committee of Ten program, set down in 1893, with another Harvard president, Charles W. Eliot, as its main sponsor.

Though James Conant did not author the 1945 Harvard report, he endorsed it fully, and it is no accident that he remained faithful to its recommendations for the next 15 years. His ideas had appeared in nascent form in an early book (*Education for a Classless Society* [1940], written with Francis Spaulding) and were laid out in considerable detail in *Education and Liberty: The Role of Schools in the Modern Democracy* (1953), in which he discussed what should be done to high schools and colleges to "accommodate" the huge increases in enrollment. His view was that colleges should be made more uniform, demanding, and restrictive, while high schools should be given a common "liberal arts" core, yet divided more broadly and intensively between college-preparatory and vocational programs, with the former being restructured around more difficult, academic-type curricula for "gifted" students. In short, one portion of secondary education should act as a gatekeeper to higher learning, and it should do this by essentially pushing the first 1 or 2 years of college education down into the high schools.

Due to his influence as an educator, Conant was sponsored by the Carnegie Corporation to do a study and analysis of the American high school and to provide concrete advice for its reform. The final report, *The American High School Today* (1959), expanded the author's platform by advising that the existing system of schools be consolidated into fewer and larger institutions. These would then command far greater resources and would be able to supply the diversity of programs demanded by a highly diverse student body. The idea, again, was not radical, but diplomatic. On the one hand, it offered a reframing of Progressive ideals. As Joel Spring (1989) describes it,

> The courses of study within the school were to be more closely geared to the future vocational destination of the student. Conant argued that the heart of the school had to be the vocational guidance person, who would measure skills and interests and arrange an educational program that would provide the student with marketable skills. . . . Conant even made the recommendation, which never became a reality, that all high school graduates be given a wallet-size record of their subjects and grades, which could be presented on demand to any employer. (p. 22)

The high school would therefore continue their paternal role (on this, he differed with the Harvard Committee), at a time when concern about "teenage behavior" was relatively omnipresent.

But Conant's larger purpose was distinctly Academist. Intelligence tests would be given to determine the ranking of academic "talent." Progressive-type schooling would be fine for lower ability students; these would all be given a basic program consisting of 4 years of English, 3 years of social studies, 1 year of mathematics, and 1 year of science (the assumption being that the latter two areas were of much greater difficulty). The top 15–20% of the pool, meanwhile, would be grouped in a college-type curriculum, which included, in addition to the English and social studies requirements, 4 years of a foreign language, 4 years of mathematics, and 3 years of "natural science" (again, biology, chemistry, and physics). This was the Committee of Ten program in unadulterated form. After high school, those students of moderate "talent" could go on to junior or technical colleges. Only those at the top would be allowed into 4-year college programs, whose overall enrollment would therefore decrease no small amount.

Conant's scheme was designed to placate both sides of a long-standing debate. To the ideal of "learning for life" he would dedicate the less able masses. To that of "standards," he would offer the "gifted." The scheme seemed to be a compromise, but it was more a ploy, aimed from the top down at the age-old separation of classes through a partitioning of means, in this case access to knowledge in a modern society. Conant sought to turn back the tide of popularization in higher learning by diverting most students to the junior college. His scheme would protect higher education by keeping it free of too much diversity, and in that way maintain its true role as the training ground for leaders in the mandarin tradition. Certainly enrollments in colleges and universities would remain well above their pre-war levels: The world was a larger, more complex place, and more leaders were required in more disciplines. Yet this was exactly the point. The "talented" had to be filtered out too.

How successful was the Conant response? At the college level, his argument for restricted access was a failure. Enrollments continued to climb. It was in the high schools that Conant's influence was felt. The place of testing, for example, as a measure of "scholastic aptitude" continued to expand. From its founding in 1947 the Educational Testing Service (also underwritten by the Carnegie Corporation) grew rapidly in influence until by the mid-1960s its various tests had become the most common screening technique for college entrance. The use of such tests, of course, had a long tradition before Conant, who simply took it up and repopularized it. In the high schools, instead of intelligence tests, a so-called PSAT (Pre-Scholastic Aptitude Test) became standard in many parts of the country. In addition, a range of new stratifications were added to segregate off and encourage more achievement-oriented students. These included Advanced Placement, Enrichment Independent Study, and Intellectually Gifted courses, all of which proved effective bribes in comparison to the so-termed Opportunity classes for the less academically inclined. But the greatest success of Conant's scheme was that, by 1965, consolidation of the high schools was a widespread reality, and the basic curriculum proposed for dividing college- and non-college-bound students had been adopted.

Finally, Conant was destined to play a critical role in American higher education from a completely different perspective, one with no small irony for his own view of science. Along with a number of other eminent scientist-statesmen of the day, for example, Vannevar Bush and in England, C. P. Snow, he helped institute a campaign within the academic community arguing that the tremendous importance of technical knowledge to contemporary society demanded it be taken up as subject of study by scholars in the humanities, especially history. In effect, the idea was meant to heal the "two cultures," to make science comprehensible in terms of its long-term development, its significance to other areas of Western culture, and (not the least) the social factors that might retard and accelerate its progress. As Larry Laudan (1990) points out, a major purpose of this effort was "not only to discover how science worked, but to communicate that understanding to scientists and the general public" (p. 55). To further this aim, Conant himself wrote a number of books for a general audience, of which *Modern Science and Modern Man* (1952) was perhaps the most well received. He also wrote textbooks for both high school and college chemistry courses, in which some of his ideas about the "incalculable importance" of scientific knowledge were given place. From these and other works, it might seem that his hope was to employ the humanities as a kind of instrument for finalizing the elevated status of science. But Conant did not look at it this way. His real ambition was to take the "meaning" of science away from scientists, whom he felt to be intellectually limited, and to turn it over to what he called "the literary tradition," the true purveyor of higher learning and social ideals.

The effect of the campaign, in any case, was immediate. By the late 1950s, history and philosophy of science departments were established or enlargened in many U.S. universities, with funding often supplied by the federal government. Conant himself undertook to organize and publish *Harvard Case Studies in the History of Science* (1959), a two-volume work, written by a range of scholars, that has remained a classic in its field (indeed, that helped *found* the field in the United States). It was while working as one of the authors on this anthology (the case of Copernicus) that the physicist-turned-historian Thomas Kuhn began to gather ideas for his *Structure of Scientific Revolutions* (1962), a book whose eventual influence would help shatter every sacred idea of science that Conant and his colleagues stood for. "History" indeed would come to teach lessons about "science"—but not the ones they had envisioned. Conant's enduring influence was to help build the "comprehensive high school," but also to lay the foundation for a comprehensive grouping of "science studies" that, before long, would destroy the old positivism.

SPUTNIK AND THE CONTINUED RISE OF EXPERTISE

No doubt Conant's success was partly due to the long-term fallout that emerged in the wake of Sputnik. It was this event, more than any other, that spurred ahead with a sense of crisis the patterns for conservative change begun earlier on.

In 1950 the National Science Foundation (NSF) was created and charged with trying to ensure U.S. leadership in all phases of scientific research and education. NSF was the brainchild of Vannevar Bush, who, as early as the mid-1940s, had recognized the new role of federal involvement in science and who therefore argued for more centralized and strategic control. His idea for increasing the store of "scientific capital" in America involved three aims: providing money and facilities for basic research; offering fellowships at the undergraduate and graduate levels; and toughening up science education in the high schools, as a recruiting ground for professional "talent." These goals were taken up in the organization's charter, along with the new goal of "securing the national defense." Indeed, any doubt about the NSF's cold war mission should be put to rest by reference to several exchanges between its director, Alan Waterman, and Congress during early appropriations hearings. On one occasion in 1955, for example, Waterman commented, "There is every reason to believe that the U.S.S.R. has already surpassed us in the production of engineers and is on the way to doing so in the production of scientists. . . . Of course one cannot go into all sources of information by which this is received. . . . [But] every evidence is that they are doing the utmost to develop their science as a source of economic strength and a strength in defense" (in Duschl, 1990, pp. 20–21).

Congress, for its part, did not yet fully agree that science was the key to winning the cold war. But 2 years later, in 1957, the launching of Sputnik changed everything. An emergency allocation for the NSF was passed almost immediately, to be used strictly for scientific education, and within 2 years nearly half of the total NSF budget (itself considerably increased) was devoted to technical training. This led, in turn, to the deep involvement of the NSF in curriculum research and design, mainly through the National Defense Education Act (1958). This act poured large amounts of new money into existing projects in this area, notably the Physical Science Study Committee or PSSC (organized in 1956) and the School Mathematics Study Group (organized in 1958), both of which had been started to find ways of teaching science and math that were more challenging, interesting, and relevant for scientific careers. Yet the production of more technical manpower was not the main or only goal. It was also, at a deeper level, to develop courses that would accelerate learning, that would make students absorb more material, at faster rates, and with greater comprehension. In a sense, the idea was not to prepare high school students for college work, but instead to make college-level work teachable to lower grades—in other words, instituting Conant's scheme in another form.

This buried intent was made clear in one of the most

popular books written on education in the 20th century, Jerome Bruner's *The Process of Education* (1960). Called "a landmark in educational theory," Bruner's thin volume summarized new ideas for reform at the primary and secondary levels, as well as some of the psychological concepts on which they were based. It spoke of "the importance of structure," it discussed "the fundamental unity" presumed to characterize each discipline, and it boldly stated "that any subject can be taught effectively in some in-tellectually honest form to any child at any stage of develop-ment" (p. 33). All of this, and much more besides, gave hope to officials, teachers, and parents that children and adolescents could be made much more successful in their intellectual labors and future careers if only the appropriate methods and courses were offered. Coming in the wake of Sputnik and at the peak of the baby boom, this type of half-misapprehension of Bruner's message was destined to have its effect.

Bruner, in fact, warned in his final pages against the rise of meritocracy. This, he said, would inflict serious harm on the democratic possibilities of learning. And yet, nearly everything else in the book seemed to head in this direction. Indeed, another of its central arguments—perhaps the most central of all—focused on a new kind of hierarchy in curriculum-building:

> Designing curricula in a way that reflects the basic structure of a field of knowledge requires the most fundamental understanding of that field. It is a task that cannot be carried out without the active participation of the ablest scholars and scientists. (p. 32)

No longer is the educational specialist, the teacher, or even the psychologist the final expert in curriculum design, especially in the sciences. Now the scientist himself must stand at the front and take the lead. Expertise with regard to education had thus been raised a final notch, elevated yet again in an old logic that sought to promote the power of those who held knowledge over those who sought to acquire it.

In nearly every effort supported by the NSF, this logic took hold. Scientists, in fact, saw the opportunity as one of control, as a chance to oversee and direct not only recruiting but the image of science taught in the schools. From the very beginning they were in charge; most were prominent researchers at major universities and had little patience with the advice of educators. The textbooks, laboratory manuals, equipment, and teacher-training programs these scholars devised were all oriented toward "science for scientists" as a particular type of inquiry. The new curriculum represented a radical change: Students were no longer to be spoonfed "facts," "principles," or the outmoded inductive view of discovery; they were now to be involved directly in actually "doing science." "Learning by doing," in fact, a phrase deriving almost directly from Dewey himself, became the phrase of the day and reveals that Progressive rhetoric was now used to reject what had become the endpoint of Progressive orthodoxy, that is, life adjustment. Demanded now, though never explicitly stated, was profession adjustment. The new curriculum appeared to be student-centered; it aimed at instructing the learner in how to work and think as a scientist. Theory was intended to play a central part as a predictor of results, one that had to be tested and evaluated. The point was to make science a process, an activity of supposition and trial, of individual judgment and "discovery."

The problem was that this image, too, comprised a type of propaganda. It was, in fact, the scientists' favorite view of their favorite subject. No doubt it constituted the image most widely held at the time among professionals, but for this reason alone it should perhaps have been suspect. First, the student was allowed no option to discover "science" as a process. This process, rather, was presented as fixed and given. Second, the student was effectively taught science-in-a-vacuum, stripped of all outside influences, utterly free of competition, of any internal and external conflicts, a "science" therefore irrelevant to such realities as funding sources, government, corporate, and military priorities, social issues, and the like. (That such "impure" priorities had been instrumental in conceiving such a curriculum which ignored their very existence was cause for no perceptions of irony; the old faith in a positivist science, free of worldly contents no matter how dedicated to worldly goals and power, was a belief still largely unchallenged at the time.) Third, this purist view gave students a false image in another sense, for it held out a more seductive portrait of scientific work, one based on *individual* exploration and the possibility for individual fame, one that therefore downplayed the reality of team effort, of "big science." Teaching such an image appealed to a student's sense of potential heroism; it linked to age-old American dreams of self-fulfillment through personal labor, of performing a grand service, contributing to a magnificent quest, even to the betterment of mankind, while simultaneously reaping the benefits of money, status, adulation, and so forth. The individual researcher, embodied in the new curriculum, was a kind of paladin subjectivity. As image, he stood a much better chance of attracting new "talent" than did the realities of team hierarchy, grant writing, and tenure battles.

The NSF-sponsored groups streamlined and toughened up the curriculum. Their distrust of the existing system made them try to produce materials that students could use on their own, that is, materials that were "teacher-proof." Many subjects and experiments formerly studied in college were shifted down to the high school, junior high school, and even grade school level. The so-called "new math," for example, sought to introduce novel techniques for arithmetic work that would allow more rapid learning and the introduction of sophisticated concepts at earlier ages, including calculus by the senior year of high school. College-level freshman classes in physics and chemistry, complete with lab work, were also reoriented and sent down to the lower grades. High school biology was restructured to involve increasing amounts of dissection, microscopic analysis, individual research projects, and so on. The order of subjects historically brought under the NSF-sponsored programs was also revealing: first physics/mathematics, later chemistry, and finally biology. As in the Harvard report under Conant, this system of priorities reflected the standard "scale of hardness" within the universe of professional science. But it was also tied to the rationale for giving physics and chemistry the lion's share of support in research generally, because these areas were supposed to have a much greater potential for enhancing America's technological "strength."

The curricula developed under the NSF programs were implemented with a speed and reach that could only be called remarkable. According to Duschl (1990), by 1965 fully 50% of all high school physics students (200,000) were using the PSSC materials, and 350,000 were employing NSF-sponsored chemistry curricula. Two years later over 3 million students were being taught by means of these and other related materials. In less than a single decade, therefore, American science education underwent a complete reformulation, mainly toward the Academist ideal. This came from the highly organized nature of the enterprises. Government money flowed in three critical directions: to curriculum design groups; to summer institutes for teacher training; and to local school districts for the exclusive purchase of new curriculum materials and equipment for science. The key link was teacher training: recruiting those teachers who would promote use of the new materials in their school systems. No tuition was charged at the summer institutes, and teachers were compensated for their time either by some form of monetary reward or by the granting of academic credit. In this way, the government acted as a kind of skillful monopoly, creating a product, a distribution system, and a captive market all at once.

Despite such a system, however, educators viewed the success as mixed, even at the time. Taking the lead on a conceptual basis, for example, Paul Goodman took the entire effort to task in saying that the goal of encouraging or producing "creative scientists" was "both pretentious and naive." The idea of "breeding" such minds, especially in school, by some pedagogic formula, seemed not merely wishful thinking but a kind of "hoax," a necessary cover for the more truthful (and practical) goal of Ph.D. preparedness. Regarding Bruner's *Process of Education*, moreover, Goodman wrote:

> To be sure, Professor Bruner and his associates do not go on to espouse democratic community. But I am afraid they will eventually find that also this is essential, for it is impossible to do creative work of any kind when the goals are pre-determined by outsiders and cannot be criticized and altered by the minds that have to do the work, even if they are youngsters. (Dewey's principle is, simply, that good teaching is that which leads the student to want to learn something more.)
>
> The compromise of the National Science Foundation on this point is rather comical. "Physical laws are not asserted; they are, it is hoped, discovered by the student." . . . That is, the student is to make

a leap of discovery to—what is already known, in a course precharted
by the Ph.D.'s at MIT. (1964, p. 54)

Goodman and others applauded the stated intentions of NSF to
teach science in a way that would bring out its own "excitement,
beauty, and intellectual satisfaction" as well as reveal its role and
importance in the everyday modern world. If handled well,
teaching could also show the power of science's "austere morality, accuracy, scrupulous respect for what occurs" (Goodman,
1964, pp. 120–121). The problem, however, was that all of
these things got short shrift in the actual programs themselves, as
written and implemented. Rather, the general drift seemed
much more to inspire an appreciation of the power of science on
the one hand, and on the other, a willingness to join in the race
for right results, high grades, quick advancement, and general
progressional aggressiveness.

Educators soon saw other problems as well. The most important of these was that the use of scientists in curriculum design
had ended up producing courses that placed the teacher in the
role of professor—that is, as pure lecturer and research tutor.
This meant that, in actual practice, styles of teaching became
more important than styles of learning. Notions of "self discovery" took a back seat to "programmed study." Deprived of recipe
instructions for finding right answers, yet still held accountable,
students ended up (naturally enough) spending a larger part of
their effort seeking to "discover" these same unwritten rules for
themselves. Where they aimed for "excitement," the programs
often delivered cynicism, something that affected teachers and
students alike. Moreover, the fact that scientists had, from the
beginning, grabbed the lead in designing the new courses and
had sometimes spoken disparagingly or even dismissively about
the role of pedagogy as a field (it was the frequent contention of
the involved scientists that only "those who do" were best
qualified to instruct "those who teach") did not at all sit well.

The effect of all this was that, by the mid-1960s, after the
scientists had departed the scene, educators began actively taking over the programs. But by this time the original motives for
their very creation, for the entire episode of curriculum reform,
were fast losing steam. The remaining history of this major effort

is aptly summarized by Dorothy Nelkin (1977) in her book *Science Textbook Controversies and the Politics of Equal Time:*

> [In the late 1960s], responding to a general social concern with "science for citizenship," the [NSF] began to shift its emphasis. . . . The declining need for scientists and engineers had coincided with an apparent disillusion with science and technology and their association with social and environmental problems. Better public understanding of science might, it was hoped, create more favorable attitudes. Again, the public schools were the vehicles for reform, and curriculum was now to be directed not only to future scientists but to future citizens in a society in which science was a crucial economic, social, and political influence. This change in perspective, however, also required new pedagogical techniques to make science more palatable to nonspecialized students. Science teaching would have to be oriented around practical problems. (p. 24)

Thus did things appear to come full circle again.

THE TYLER RATIONALE: A BRIEF NOTE

The continued rise of expertise in education also found expression throughout the field of pedagogy. This movement was related, for example, to a new stage of infusing curriculum building with "theory" and "method," founded on the ideas of Ralph Tyler, a professor of education at the University of Chicago. Tyler's program was itself first put forth as curriculum material, in the form of an upper-level course syllabus entitled *Basic Principles of Curriculum and Instruction* (1950). It was perhaps fitting that a new rationale for curriculum making would emerge from a situation in which education was studying itself.

Tyler's rationale, in fact, was not entirely new. It might best be described as a more systematic handling of older concepts, in particular those surrounding the idea of "objectives." What was new could be found in the emphasis given to theory: Though never really admitted, the use of "science" as a model here is clear. Theory was to become the unifying principle or directive that would make sense of required data from three sources: study of learners, analysis of contemporary living, and suggestions for content made by scholars or specialists in a particular field.

When such empirical information was properly "screened," a curriculum could be built, step by step, in a process that involved the successive answering of four fundamental questions: What educational purposes should the school attain? What experiences can it offer to achieve these? How can these experiences be organized? How can one evaluate success?

The reasoning was instrumental, deductive. It viewed the school as "performing a function," not as a Deweyan community or a social/intellectual situation, but instead as a kind of testing ground for yielding results that could then be measured. Tyler's view was that "education is a process of changing behavior patterns of people" (1950, p. 4). Students, as learners, should be studied in terms of how they act and perform in their daily lives. Various norms should be established regarding such activities as reading, speaking, interests, and so forth. These would yield "objectives" for "broadening and deepening" a student's participation in such norms: The effort should be "to identify needed changes in behavior patterns . . . which the educational institution should seek to produce" (pp. 4–5). School is therefore an instrument not merely of control or guidance but of manipulation for beneficial ends. Students, in the end, are less important as individuals than is their behavior as necessary "data," as an indicator of design success or failure. Indeed, one might note that structuring the curriculum around fixed, measurable objectives immediately places education in such a functional frame, making it a process that must be imposed, observed, and tracked in terms of changes in some initial (lower) state. As Kliebard (1992) remarks,

> One reason for the success of the Tyler rationale is its very rationality. In some form, it will always stand as the model of curriculum development for those who conceive of the curriculum as a complex machinery for transforming the crude raw material that children bring with them to school into a finished and useful product. (p. 164)

One should note that Tyler's general scheme came to be adopted as doctrine by most educators concerned with secondary schooling not only during the 1950s, but throughout the 1960s, 1970s, and 1980s. The fourfold path to enlightened curriculum building remains largely in place today. Its basic formula, behaviorist in spirit, was enthusiastically approved during the 1970s in particular, when Skinnerian notions of behavior modification were taken up in the schools. During the earlier postwar era, meanwhile, it coincided with the rise of the test industry, mentioned above, and grew into a movement for defining the "objectives" that all schools should pursue, as in Benjamin Bloom's (1956) classic and much-used *Taxonomy of Educational Objectives*.[3]

Finally, the popularity of Tyler's scheme helped finalize other aspects in American schooling. On the one hand, it gave "education" a firm and usable *meaning*: To the question of what education was supposed to be or to do, there were now clear answers that could be decided upon, set down, and codified in action. Beyond this, it also put the educational expert in the driver's seat, once and for all. By making curriculum design dependent on theory, it made the school subservient to the theorist, he or she who—even as a "hired gun"—was now to decide what direction teaching and study might take. By stressing the need for "assessment" as the endpoint of the process, moreover, the scheme ended up by confusing "evaluation" with testing, that is, with quick, efficient, technofix methods of categorizing students in terms of their recitational (technological) ability to perform. Once established, this confusion became a self-fulfilling prophecy: Test scores rose to become "objective" measures of success for the school, the teacher, and the student. Testing was thus given power to define "education" too in terms of an extremely primitive model of labor based on senders and receivers, a fact that had no small part in the general rise of pedagogic behaviorism during the 1970s.

The ample employment of Tylerian ideas as institutional doctrine, beginning in the 1950s, helped finalize a profound division of territory in educational psychology. This had existed before, since the early part of the century, but it now gained a distinct form. The tradition of child study, now the province of developmental psychology, and its concern with mental growth, came to be applied exclusively to primary school. For older children, behaviorism came to dominate. The split was important; it showed that different definitions, indeed, different sciences, were applied to each realm. In the primary grades, instruction was aimed at the child; above this, there was a transformation into the student. Learning as an affair of individual experience and growth—the old feeling associated with "the natural"—was increasingly reserved for young children, those "apprentices to life" (Goethe) most innocent of all knowledge. What then took over, however, was "education," the institutional process of training and control. The age of innocence died early, therefore. Society, in the form of purposes and objectives and tests that cared little for the living person, came crashing in like a cold wave. An interesting closure resulted. Bonds of intimacy and tutorship existed only at the beginning, where learning began, and at the end, where it ended, in the apprenticeship of the graduate school. Everywhere in between another "science" came to rule.

New Voices and Old Limits: Reformism in the Late 1960s

In the single decade between 1965 and 1975 American education entered and abruptly left a new period of major Reformism,
aimed at all levels of schooling. Entrance had everything to do
with political and social events. During the 1960s the civil
rights movement became a central force on the domestic scene,
prompting reform and a deep questioning of existing attitudes,
behaviors, and social structures. This helped launch, in turn,
the Free Speech Movement at the University of California's
Berkeley Campus, and then Students for a Democratic Society
(SDS) and the entire protest movement of the late 1960s and
early 1970s, which carried both demands for change and political tactics into the very heart of academia proper.[1] The cold war,
meanwhile, whose rhetoric of threat and fear was already under
heavy criticism, culminated in the Vietnam War, an event that
greatly elevated the scale and intensity of protest. All this
occurred finally, in a period characterized by economic prosperity and by swelling enrollments in secondary and tertiary educational institutions as children of the baby boom reached their
teens. The effect of these last two factors was essential: The
economy allowed government to respond with a host of new
social welfare programs (even while continuing to escalate the
war); and the new student masses, once mobilized, constituted a
formidible political force that could hardly be ignored.

The combination encouraged an episode of widespread
radicalism that was destined to come back to the schools in
which it had been nurtured. As in the past, education was again

viewed as the provenance of most social realities, but now in negative terms that directly implied the liberating orientation that reform should take. Schools and colleges were perceived as an indoctrinating force, part of a "system" that had become the source for a society rife with authoritarian, racist, sexist, and dehumanizing aspects. Academic training in the traditional disciplines was argued to produce an overly rational, denaturalized mentality, disconnected from emotional needs; the academy was a "machine" that stamped out human beings who then submitted mindlessly to the service of "runaway technology." The idea of the university as a place of nonpartisan scholarship and ecumenical service to society—the old "Wisconsin idea"—fell under attack for having become a mask to the realities of financial dependence and philosophic loyalty with regard to "the military-industrial complex" (a term used by President Eisenhower, in warning, during his farewell speech of 1960). Moreover, the experience at Berkeley had shown beyond all doubt that universities were anything but democratic organizations in practice, that there were strict limits placed on what students could say and write, that they were in fact denied many basic civil rights within the gates of "higher learning." Fifty years after Ellwood Cubberley and the efficiency movement, the image of the pupil as "raw material" and the college as a "machine" was brought back as an icon of outrage, tyranny, and rebellion.

The Politics of Change and
the Role of Science

At every level of schooling, therefore, the solutions seemed relatively clear. What had to occur was a release from the past, an injection of new freedoms.

Federally sponsored reforms were the earliest, ironically, since they were tied to pressures emerging out of the civil rights movement during the early to mid-1960s. Now in the hands of the Democrats, government patronage and policy performed an abrupt turnaround from its Cold War Academism of only a few years before. It was not that "intellectual skills" were now viewed as irrelevant; on the contrary, they remained central. But the goal to which they should be devoted changed, shifting from "technical manpower" to individual equality and opportunity. As the primary origin for these things, education seemed to promise the solution to nearly every pressing social problem of the day. By a kind of beneficial domino effect, equal education would help cure such things as the "maldistribution of skills" and racism (which began in childhood), and from there would go on to alleviate unemployment, poverty, crime, broken families, and urban decay. The logic was close to utopian, and hung on a series of assumed connections between all these phenomena. It revealed again, however, the enormously raised expectations that had come to settle on the school since the turn of the century. Of course, the federal motive was not without other Practicalist intentions: By uplifting the poor, after all, and ending an important source of long-term internal weakness, American democracy would become better able to stand firm as the leader of the free world. But this was secondary to the domestic agenda itself, pushed ahead by the 88th and 89th "Education Congress" under Lyndon B. Johnson.

Legislation was enacted to support schools at every level; billions of dollars were appropriated for the tasks ahead. New programs such as Head Start were begun to help mitigate the "cycle of poverty." Job training, vocational and continuing education, and other compensatory federal efforts were set up under the Economic Opportunity Act (1964) and other new mandates. Passage of the 1965 Elementary and Secondary Education Act provided major support for research on learning, instruction, and curriculum development. Its task force began with the premise that "We must overhaul American education . . . shed outworn educational practice, dismantle outmoded educational facilities, and create new and better learning environments" (in Spring, 1989, p. 220). The terms were identical to those of the Sputnik reformers a decade before, who were now cast as outmoded. The call for "overhaul" was once again a demand not merely for change but for taking control. It meant that a new administration, with its proclamations of "the Great Society," would try to remake education (and thus society) in its own image of beneficial intent of what America could and should be.

That the government's role was bound up with ideas of research, social and psychological, reveals that science remained no less central to ideas of official reform than before. This "science" was very different from that of the cold war "brains for freedom" initiatives. It was not centered on the natural sciences or on training in technical fields for the sake of economic and technological power. It was instead about using expert knowledge in the social sciences to better social conditions. It was a return to the Progressive idea of "science for human advancement."

Indeed, it was a widespread idea at the time. Both within and without government, the multifold links between ideas of "science" and those of "education" asserted themselves in varying forms. The child in general again became a subject of much concern, the site both of wounded references and of ambitious renewal. As in the 1890s, distrust was leveled at the school as a place where children were being taught in ways that denied their innate growth, their curiosity and individuality. Child study again became a subject of much popular and expert attention— now relabeled as the "science of childhood"—with the ideas of those such as Jean Piaget, D. W. Winnicott, and A. S. Neill, among others, attracting much notice.

One part of the new child study initiative carried forward Bruners' idea that children were capable of learning a great deal at an early age. This theory gained great support through a number of government-funded studies during the mid-1960s on the intellectual capacities and development of underprivileged children. Seeking to directly overturn ingrained notions of racial and ethnic inferiority, educational psychologists elaborated the concept of the "competent child." They emphasized the effects of environment on IQ, and they spoke of "sensory deprivation." But more than this, as expressed in perhaps the most influential book of the day, Benjamin Bloom's (1964) *Stability and Change in Human Characteristics*, they layed stress on the notion that "general intelligence appears to develop as much from conception to age four as it does during the fourteen years from age four to eighteen" (pp. 207–208). The conclusion to be derived from this was a simple one: Children who did poorly in school were inadequately prepared at a younger age. "Early childhood education" therefore became an idea translated on the one hand into a warrant for government programs such as Head Start and television programs such as "Sesame Street," and on the other hand, for toughening up the curriculum in elementary schools (under the aegis of making it "more challenging"), even to the point of pushing the three R's down to the kindergarten level.

Meanwhile, many teachers and parents, inspired by books such as A. S. Neill's *Summerhill* (1960), H. Kohl's *The Open Classroom* (1969), Ivan Illich's *Deschooling Society* (1970), and Allen Graubard's *Free the Children* (1972), sought to escape or avoid the existing system altogether by founding their own experimental schools in the Deweyan tradition, or to seek out and support other alternative choices such as the Montessori schools. Historically, the sentiment which gathered in this direction may well have begun with the popularization of Neill's work in particular by the mid-1960s. "The function of the child," Neill wrote, sounding the paean, "is to live his own life—not the life that his anxious parents think he should live, nor a life according to the purpose of the educator who thinks he knows what is best. All this interference and guidance on the part of adults only produces a generation of robots" (1960, p. 2). School, then, as it had come to be, was tantamount to incarceration, for everyone concerned: "The classroom walls and the prisonlike buildings narrow the teacher's outlook and prevent him from seeing the true essentials of education" (p. 28), which, in the end, came down to providing children "the chance to grow in freedom" (p. 91).

During the early 1970's, this disaffection with public schools blossomed into the widespread by short-lived Free School Movement, with its child-centered philosophy. Indeed, the entire experimental tradition in American education, always active around the margins, moved into the mainstream at this time. As implied, schools were often compared to prisons, and the need for new and liberating connections between "real life," a nurturing environment, and learning were loudly proclaimed on every side. Much reinventing of wheels took place. Dewey, Parker, and other reformers were rarely read or ressurrected. The point, one might say, was to forget history, to indulge the enthusiasms of "revolution." Yet it is striking overall how much of what had been thought, said, and done by these men more than 70 years earlier was now rediscovered and reapplied during the new period of Reformism. This repetition suggests, perhaps most of all, how little things had really changed in public schooling at a fundamental level over the intervening decades, and therefore how likely, at some point, a resurgence of similar Deweyan criticisms would emerge and quickly lead to similar diagnoses, hopes, and plans for change.

The Head Start curriculum, for example, as designated by new research, was comprehensive, proclaimed as a break from the past. Aimed at prekindergarten children, it involved many of Dewey's ideas: activities concerning health awareness, natural materials, alphabet and number play, storytelling, and cooperative game playing. It was not an academic program but rather one that intended to provide "enhancing" materials to make up for what might be missing in the home. It sought to include parents directly, to bring the family into the school and the school into the family. Its conscious design was to offer "experiences in learning," to create "an interest in learning," or even to teach "learning how to learn." Such phrases, however, reveal a deeper mixture of Deweyan ideals with those of a more manipulative stripe. The prime motive, after all, was to prepare children for mainstream schooling, to help compensate at an early age for the competitive nature of educational experience later on.[2]

For the high schools, in contrast, another type of official "science" came into play. In addition to ideas for reform coming from students and educators directly, a new round of expert policy studies were initiated. Since the days of Conant's (1959) *The American High School Today,* sponsored by the Carnegie Corporation, this mode of taking control over the development of a new philosophy for change had become fixed in the outlook of many foundations, as well as the government. The idea was by no means new; but it had never enjoyed such success among the public, academia, and federal officials until the Conant report. During the late 1960s it became a means of enlisting professional experts from various fields to displace the old idea of "service" into a new context. Attacked on the one side, "the Wisconsin idea" was held up and invigorated on another because its supporters were again politically powerful. Indeed, the various blue-ribbon panels and task forces assembled for each new round of reports, aimed at getting the "objective facts" out in the open for policy use, resembled nothing so much as an ad hoc version of Walter Lippman's old think-tank concept. The result, in any case, was a series of reports on the condition of secondary schooling that agreed in all their basic tenets. As Cremin (1989) notes, the turnaround from a mere decade before is striking:

> The American high school [was now] an institution victimized by its own success: the closer it came to achieving universality, the larger, the less responsive, and the more isolated . . . it became, walling adolescents off from other segments of society, organizing them into rigidly defined age groups, and locking them into tight and inflexible academic programs. . . . High schools needed to be reduced in size so that students could be dealt with as individuals; curricular options needed to be broadened and diversified; . . . students needed to be empowered to make more of their own decisions. (pp. 26–27)

That the experts all seemed to agree with the tenor of the times did not in the least disturb anyone's faith in their qualifications. Nor, in the 1980s, when they argued that these very recommendations had helped create a "rising tide of mediocrity," did any important suspicion arise (except among historians, perhaps) that the games of influence being played were by then part of education itself.

Perhaps more than any other level of education, however, it was the university that drew the most acrid fire and activism for reform. Standards for admission, teaching methods, and the content of many courses were attacked for being out of touch with life beyond the ivory gates, or as authoritarian, or else as biased in favor of certain socially ingrained prejudices, such as those having to do with race, gender, and class. Here "science" was less an idea for diagnosis than a target for change itself. Some groups saw it as the essence of domination, of a materialist civilization, bent on profit through the destruction of nature and the natural. Others inspired by romantic sentiment viewed it as cold and analytical, antagonistic to human life, to feeling, to peace, to health. Still others discovered in it a projection of the white, male view onto the material world. A good deal of

criticism, from students and intellectuals alike, focused on perceived collusion between the research community, the government, and the corporations, impugning the role of science and
scientists in the production of military weaponry, corporate
expansionism, and environmental pollution. This last, in particular, led to widespread protest and to demands for reorienta-
tion, but to little overall change. Existing structures and institutional relationships between technical knowledge, research,
and higher education were by this time much too critical to the
financial existence of the university itself; indeed, they had been
so all along and could not be overturned within the brief period
spanned by the student movement.

"Science," in any case, had been made a subject of deep
controversy, a more solid object of blame and disenchantment
than ever before. As always, what exactly was meant when one
or another group invoked "science" could not often be made
clear. Relevant perceptions tended to accept the traditional
status of "science" as a grand reified entity, but beyond this one
saw any number of vague definitions at play. For some the evil
was a state of mind, a way of looking at the world (as a dead
object, an exploitable resource, a despiritualized entity); for
others it was a set of power relationships at the heart of capitalism; still others, however, spoke of "science" as a matter of
discovering and unleashing dangerous forces, or else as a collection of cold facts that suppressed the emotional, nutritive side to
human life. Finally, more than a few intellectuals rolled all such
attitudes up into a single, tangled ball of rejection. Yet all this
was perhaps less important than the general feeling that something had gone seriously wrong in the longtime American love
affair with this enterprise, science, whose final aims could no
longer be assumed to lie in the realm of human benefit and
democratic progress.

One should not suppose, however, that "the scientific" was
universally condemned. Most students majoring in the natural
sciences did not even take part in the protest movement, for
they were too intent upon pursuing academic and research careers. And even for those who did, and who undertook to create
within the college itself alternative "universities without
walls"—an equivalent, one might say, to student lyceums—the

language most often adopted as a banner for such efforts was not a political or philosophic one, but instead one derived from scientific work, "experiment," as in experimental colleges, experimental curricula, and learning experiments. Such learning, as a type of cooperative exploration, so reminiscent of Dewey and of Jefferson, was consistently revived during this period. In some sense, the idea of "the American experiment," so vital and important to the early Republic, was recast in the hope for a new, liberating type of education. Thus, even as an idea, as a source of imagery, "science" remained larger and more useful than its damaged reputation.

Most of this activity took place between 1965 and 1974. Its activist breadth was striking, and so was its diversity in terms of claims and ambitions. In the lower grades a number of schools toyed with open classroom ideas, at least briefly. This involved dividing the space of the room into various designated "learning stations," with relevant materials available for students to move from station to station during the day, guided by their own interests. In many high schools and in most colleges, meanwhile, the curriculum was both relaxed and broadened; core requirements were eliminated or greatly scaled back, while new electives oriented toward study and discussion of social problems were created. Some experiments with the hour- or credit-system were tried out, involving minicourses, interdisciplinary majors, independant study, and work-study arrangements. Here, too, many classrooms for a time became more "open," with desks arranged in circles, meetings held outdoors or off-campus, and exams less based on recitational (true/false; multiple guess) questioning. Teachers and professors often tried to encourage more in-class dialogue and were frequently more lenient in their grading. The general atmosphere was far more informal than it had ever been.

Yet, almost as suddenly as they had begun, the great majority of these experiments came to an end, again for political and economic reasons. Most major institutional changes, after all, had been dependent either on direct government support or on continued high-level funding in general. The schools had not been flush, exactly; but they were certainly secure, well supported and attended to, and they conceived their future as being undaunted. This, then, gave many of them the boldness to try out new ideas, to respond to the new attitudes spreading through American society. Yet what this showed was that reforms of the past had not freed education from overarching control by "outside" forces; in fact, they had dramatically increased it. Fragile in their confidence, the schools were poised for a serious setback with even a small loss of support. Such loss came, and it was not small. By 1974 the Vietnam War was being rapidly scaled back and the Arab oil embargo was in progress. The economy fell abruptly and abysmally into an episode of unemployment, recession, and then inflation. The enormous resources that had been poured into rethinking educational ideas and practices dried up almost overnight. Within a few brief years, the coalition of interests that had battled for a "new public education" fell into confusion, broke apart, and dissolved into a scattering of local efforts. Popular and federal priorities skidded away from commitments made in this area throughout the mid- to late 1970s, and what remained by 1980 was killed outright by the conservative policies of the Reagan administration.

Even by the early 1970s right-wing interests had begun to assert their power. The backlash, in fact, took place on several levels. By 1971 it was becoming apparent that the country was headed for economic hard times; unemployment and, more specifically, underemployment among recent recipients of graduate degrees was a well-recognized fact. Ph.D.'s in literature and philosophy were driving cabs; research posts in all the hard sciences were being reduced just at the time when the greatest outpouring of qualified applicants was underway. The result was a major federal program earmarked as "career-education," with over $100 million of discretionary money being devoted to pilot programs by 1973. Such education, needless to say, had little patience for "experiment." It aimed at merging traditional academic training with a vocational focus—in short, the basic program that had existed prior to the late 1960s.

Beyond this, conservatives, feeling legitimized by the reelection of Richard Nixon, rose up against all student-centered ideas of learning, which they said had "toyed" irresponsibly with the institutional apparatus and had promoted cultural and moral relativism in the minds of the young. College campuses were targeted as the source of sexual permissiveness and a host of other evils. In seeming echo of the agrarian movement 75 years earlier, a new wave of fundamentalist opposition broke out, attacking secular, antipatriotic, and ungodly attitudes in the schools, prompting book-burning campaigns in many states, and arguing for the equal-time installment of Creationism as a "scientific" alternative to the existing biological curriculum. This movement, in turn, was quickly followed by a more broad back-to-basics movement, whose overall hope was to restore lost discipline and to make learning at all levels more accountable to conventional forms of performance. In precollege classrooms, especially, as John Goodlad (1984) has noted, growing "pressures were exerted on teachers and, ultimately, students to do better, particularly in the 'basic subjects' " (p. 7). Highly organized, backed both by corporate interests and by conservatives in the school system itself, the movement led local schoolboards to quickly reinstated the old curriculum wherever it had been suspended or amended.

Within this movement, too, "science" came to be wielded as a pointed implement in several ways. One way involved the use of "objective" testing. Many states, for example, instituted annual standardized exams and made teachers responsible for raising scores (their salary increases depended on it). Students, on the other hand, were now often forced to take proficiency exams in order to advance or to graduate. Tests were also used to bring back or reinforce tracking as a common practice, despite the enormous amount of material indicating its disadvantages. A noted decline in SAT scores, evident for the period 1966–1975, was blamed on the teachers for being too "permissive," and on the schools for being too "easy on standards."[3] Somehow, as Goodlad (1984) points out, such results were assumed to show that the entire *condition* of the schools was degenerating, that Reformism had weakened and polluted it to such a degree that only immediate authoritarian measures could save the American educational system from total collapse. Still more, such "scientific measures" were claimed to be "proof" that the influx of minority students had "lowered the average," and thus that they were "not ready" for college. Academism reasserted itself in the idea that too much popularization meant too little quality, that higher learning in particular had to be protected from too much democracy. One should note the parallels here with such early promoters of testing such as Thorndike and Terman. Minority students were in a certain sense seen as recent immigrants in the world of higher education.

In the lower grades, the idea of the "competent child" gave rise to the concept of "competency-based instruction." This theory was founded not on Piagetian concepts but on behaviorist ones, posing the teacher as a type of efficiency engineer, direct-ing the process of learning toward some predefined objective. The view was technological (behaviorism in general is a tech-nological response to the question of human action), and it involved mechanistic procedures. Children were now to be taught in terms of fixed course units, quizzed and graded in stepwise fashion, and made to follow detailed rules of conduct, both in and out of class. In conjunction with ability grouping, statistical approaches to diagnosing the "health" or "illness" of any individual school became popular, eliminating the need for principals and administrators to keep in direct human contact with teachers. Child study as a mode of thinking about curricu-lum issues, meanwhile, was again sent down to the earliest grades, or, especially toward the end of the 1970s (when daycare issues began to surface), to the preschool years. Everywhere else, the Tyler rationales (see Chapter 8) were brought back in force, along with the curriculum-building procedures they incarnated: listing objectives, describing processes, and measuring results.

This rapid adoption of careerism, on the one hand, and behaviorism, on the other, reflected the great loss of idealism that had occurred and its replacement by an older system of belief in mechanical solutions to the demand for "output." As Cuban (1984) indicates, "Compared to earlier reform efforts, the brevity of interest in informal schooling was astonishing. . . . [B]y the mid-1970s, concerns for basic skills, test scores . . . and minimum competencies had replaced open classrooms on the agenda of school boards, superintendents, principals, and teach-ers" (p. 151). The old structures of public education, still present despite surface changes, were relegitimized and enforced by such an agenda.

During the entire period of the late 1960s and 1970s the "science" taken up by mainstream liberal and conservative interests alike was of the traditional positivist type. Science meant the Academist ideal of final truth. Even if interpretations were arguable, "data" and "method" were not, and these things were returned to, time and again, as the ground-zero evidence of the need for change in either direction.

But beyond the pale of official claims making, change was coming to "science" too, and in no small way. Both among the general public and within academic scholarship itself it was soon to become a new feeding ground for skepticism and for inquiries of disbelief.

The Age of Lost Innocence

Public images of science and technology changed dramatically in the late 1960s. This was partly due to continuing fears about nuclear war and to new revelations regarding the degree to which research was linked to the arms race. More important, however, were issues such as pollution, toxic waste, nuclear power, carcinogens, DNA research, and medical malpractice, all of which threatened to touch people's lives in direct and immediate ways, most of which came to link technical knowledge with overarching and unaccountable authority. The controversies that came to surround these issues were followed closely in the media. In many cases the scientists who were involved, ill-prepared as they were for the requirements of public scrutiny, appeared arrogant, elitist, and dismissive of such hallowed American values as "the people's right to know." A feeling grew up that science and technology had been left unexamined for too long; they needed to be taken more in hand, guided by public scrutiny.

Again, however, distrust and skepticism were not the final word. They were still to be mixed with belief in the potential good uses and corrected heroism of science. To the extent that the siting of nuclear power plants excited opposition, the landing on the moon in 1969 was everywhere hailed as "a miracle of modern technology." While doctors were increasingly seen as detached and untrustworthy, President Nixon's "War on Cancer" was greeted as a promise to defeat disease with more science, and succeeded in vastly enlargening the National Institute of Health's (NIH) budget and thereby bringing medical research under the federal-grant system.

On the other hand, science received edification of a very different kind, with the return of a much older sensibility. The Reformist period, after all, marked a renewed connection between science and pastoralism—American science and American nature—as put forth and embodied by the environmental movement. The idea of an "American nature," of course, was not made explicit under the same nationalisms as in the 19th century. It remained mostly buried in references to an unspoiled, native magnificence. A difference today is that direct contact is no longer enough by itself; the mission to witness America and learn from its beauties must be combined with a movement to save it. This idea is evident in notions such as "our vanishing wilderness," "endangered domestic species," and "restoring balance to our forests." Within this discourse of "our" is a quality of piety, a displaced patriotism. The concern for nature cannot be divided from a concern for America—an America not of people but of land, animal life, trees, and waters, of myriad organic vulnerabilities. This nature, and therefore this America (which must be saved from itself), are also inseparable from science, which both provides a modern definition of the problems involved, through ecology, geology, zoology, and so on, and the possibility of protecting "our natural heritage" through knowledge. Science is again connected to moral values through aesthetic and spiritual priorities, through ideas of beauty, purity, and nativeness—all related to a concept of "America" as a unique idyll and an origin of national pastoralism.[4]

Both the advent of public controversies over science and the environmental movement had major effects upon the curriculum of higher learning. One result was to contribute new experts to deal with the new realms of social concern. Programs on science, technology, and society (STS), bioethics, science policy, and other crossdisciplinary areas were begun in many schools, drawing academics from a range of different fields, including the natural sciences. By the end of the 1970s such programs had become well-recognized disciplines in their own right and were supplying trained personnel to government and industry, as well as back to academia. Philosophically, such study marked one part of a new trend within the university of making science accountable for its effects in culture.

New fields were also begun within the sciences. Those related to environmentalism, for example, tended to straddle the borders between science, engineering, and "social conscience," and were in line with the shift to careerism. The teaching of courses like systems ecology, pollution chemistry, and wetlands management required a different approach than strict research apprenticeship. Many of the techniques pioneered by Progressive education—field trips, site visitation, team learning, and use of real-world problems—were employed or refined. The result, again, was a new cadre of professionals for the government and also for the new and burgeoning environmental industries.

These effects within the curriculum were significant. They were corrective interventions, one might say, that adopted the idea of service to larger public goals. They did not have much to say about the actual contents of technical knowledge: They spoke only to its uses. If science policy, for example, took the notion of science-as-national-destiny for a given, it had little notion of what this meant beyond certain obvious forms of Practicalist power.

The more profound intellectual reforms took place elsewhere. They came, in fact, from a series of efforts that took "science" as a subject of critical thought and that ended in attempts to dismantle the positivist hegemony. From the mid-1960s onward, science and scientific work became a kind of frontier subject for a multitude of different research possibilities. The idea of objectivity and final knowledge was taken apart, piece by piece, by being made in effect the object of disbelief and analysis in one field after another: in history and philosophy first of all, then in sociology, psychology, economics, even anthropology, semiotics, and literary (discourse) analysis by the early 1980s. It was as if the social sciences were having their oedipal revenge.

Kuhn's "Revolution" and the New Insider Image of Science

No doubt the beginning of this trend in the United States can be traced to the influence of Thomas Kuhn's (1962) *Structure of Scientific Revolutions*. Kuhn's book, which had a transformational effect on the perception of scientific advance throughout academia (and thus on how it might be taught), opened with what was surely a warning, even an indictment:

> History, if viewed as a repository for more than anecdote or chronology, could produce a decisive transformation in the image of science by which we are now possessed. That image has previously been drawn . . . mainly from the study of finished scientific achievements as these are recorded in the classics and, more recently, in the textbooks. . . . Inevitably, however, the aim of such books is persuasive and pedagogic; a concept of science drawn from them is no more likely to fit the enterprise that produced them than an image of a national culture drawn from a tourist brochure or a language text. This essay attempts to show that we have been misled by them in fundamental ways.

Kuhn went on to dismantle the image of science held by nearly all academics of the day, an image of step-by-step advance, punctuated by individual geniuses, purely self-driven and purely an activity of mind *(theoria)*. This the author replaced with a modified "internalist" model, Hegelian for the most part, based on the concept of "paradigms." These were unifying structures of thought that served as gospel, being maintained not because of their truth value but because of unreflective loyalty, intellectual conservatism, and propaganda-like training. Paradigms sooner or later generated problems they could not solve or explain, at which point a "crisis" would begin to occur. This built to the point where full-scale "revolutions" ensued, whose advent was marked not by a period of celebratory conversion but rather by intense conflict, rejection, and "psycho-social upheaval" (Porter, 1990, p. 38).

Kuhn's theory itself was considered "revolutionary" for several reasons. First, it introduced discontinuity into the idea of scientific advance (this idea was already established in continental philosophy but was virtually unknown in America).

Second, it portrayed the inner workings of science as a matter of human beings engaged in devotions that were political and sometimes reactionary by nature, that is, concerned with extending an intellectual status quo. Third, Kuhn implied that no paradigm could be judged better than any other on the basis of some exterior objective standard of truth. This was the concept that most enraged or surprised (or delighted) the existing community of scholars.[5] Many accused the author of relativism; most understood that such "incommensurability" would allow factors external to science itself—political, economic, cultural in the broad sense—to enter in and possibly affect the contour or choice of a theory. (Kuhn himself meanwhile, tried to make peace while downplaying these aspects of his theory.)

Finally, Kuhn's book discussed scientific training. This he posed in terms that appear almost deliberately intended to incite anguish as "a narrow and rigid education, probably more so than any other except perhaps in orthodox theology." Science relied upon the textbook as Holy Writ; it did not ask its students to read past works, and it did not ask them to know anything about the detailed development of their discipline. As a result, instead of being experts on the broader nature of their work, scientists were most often "little better than laymen" in this area, more frequently the victims of images archaic in content and often prejudicial in meaning. With regard to their own research training, moreover, Kuhn had this to say:

> The process of learning by finger exercise . . . continues throughout the process of professional initiation. As the student proceeds from his freshman course to and through his doctoral dissertation, the problems assigned to him become more complex and less completely precedented. But they continue to be closely modeled on previous achievements *as are the problems that normally occupy him during his subsequent independent scientific career.* (1962, p. 47; emphasis added)

Apprenticeship to an existing system was therefore the mold for all further work. Believing themselves heroes, scientists were more akin to intellectual hoplites. And yet, despite all this, it was the scientific community that would have the last historical word. Within a decade Kuhn's ideas had been taken up nearly everywhere, at least in selective fashion. The theory satisfied two important appetites: First, it offered a rational, almost mechanistic story of scientific advance; second, and more importantly, it neatly fit a belief that the "big science" wrought by full-scale federal support was revealing "an endless frontier," that an age of permanent revolution had begun in which every scientist could take part. It played upon the awareness of scientists that their work was, in fact, engaged in certain large-scale struggles (e.g., between old and new), yet it also embodied a psychology of heroic change that began with individuals. According to Kuhn, that is, a scientist could obey all the rules of professionalism yet still be revolutionary in his work and thought. He could be a member in good standing and a radical innovator at the same time, a compliant follower and a pioneer. The idea of "revolution," to be sure, had been attributed to science before. Indeed, the notion of technical knowledge having brought a "revolution in daily living" had become a fixture of public language by the 1920s and had found its scholastic echo in analyses of "the scientific revolution" of the 17th century. But Kuhn helped give the idea a new twist for American intellectuals. Not only did he make it intrinsic to technical knowledge; he seemed to imply that a "revolution" could begin at almost any time, given a sufficient quantity of novel data or theorizing.[6]

The uptake of Kuhnian language must be seen against another background. By the late 1960s, research growth had become viewed by most scientists as an irreversible right, essential to the advance of knowledge and the welfare and strength of the nation. The 1970s, however, saw science under serious public attack, and, as the country went into recession, under the threat of major economic cutbacks. There was a need to restore faith and to keep the claims of research as high as possible. By the end of the 1970s, therefore, it was nearly a reflex for every major new hypothesis, discovery, or technology to be dubbed with the "revolutionary" adjective. Such had become the professional image among scientists. It entered the textbooks and became part of the atmosphere of learning and doing science at the upper levels. It also became the language of public speaking and was soon adopted in full by the national media, thus adding important symbolic capital to the public view of scientific work. The term "paradigm," filtered through a great deal of discussion in the humanities, not the least in literary criticism (which itself had become partly scientized, beginning in the 1950s era of Academism, in its love of "theory"). As with the sciences, Kuhn's theory provided a means by which fields of scholarly endeavor could absorb the larger cultural atmosphere of excitement and privilege being granted ideas of change and overturn. It offered a tool both for casting a historical frame around a discipline and valorizing it. In an ironic, but meaningful, twist, Kuhn's book proved its own point a dozen times over about "externalist" factors and their potential effect upon scholarship. Indeed, in its great utility and hyper-influence, it demonstrated this point far beyond anything ever imagined by the author (whose claims regarding the matter were rather tentative in any case). Kuhn's work was, one might say, a critical resource for the historical moment—but only because the moment saw fit to make it so.

In the end, an older insider image of science—pivoting on "disinterest" and "removal"—was quickly replaced by a new one, no less heroic in tenor, yet better suited to the political needs of the age. At a time when true social change seemed in the making, when resistance to established institutions rose on many sides, establishment science might be said to have stolen the language of the moment. The self-image of scientists was at stake; the other face of "big science," after all, was that it had made technical work a mass form of labor, dependent on government priorities most of all. "Revolution" arrived to actually trump these realities on their own terms, to promote a new "science" through appeals to competition and unending "breakthroughs."

SCIENCE STUDIES AND THEIR LIMITS

Though a latecomer on the Western scene of positivist debunking, Kuhn's book was nonetheless the shot heard round the ivory tower in America. For the humanities and social sciences, it also became a tool, but for purposes of scholarly expansion. By suggesting the importance of sociopsychological factors in scientific work, it helped urge scholars to turn away from the former concentration on pure meaning and focus instead on how knowledge came to be made.

Inquiry in various fields began to follow the social history of science, the development of its institutions, its ideological conflicts, the influences of culture upon it. The influence of Marxist writings (e.g., from the Frankfurt school) and of unity by such French philosophers as Gaston Bachelard and Michel Foucault, meanwhile, encouraged new thinking about the cognitive content of scientific knowledge, and especially about how such content incorporates the historical moment in which it is embedded. As a result, giants such as Galileo or Boyle came to be studied in terms of the beliefs they absorbed and the social conflicts they reveal in their work. Their classic texts were reread as objects of ideologic analysis, and, beginning somewhat later, in terms of their narrative strategies or "literary technologies." The question of how a new fact comes to exist was pursued anthropologically into the very laboratory itself, where the detailed codes and actions of "the scientific tribe" were observed, catalogued, and interpreted.[7]

By the late 1970s the academic view of "science" had come to broadly mirror that of the general public. Perceptions regarding the political and social context of research had become a working base for inquiry in a dozen different areas. Yet the new "externalist" scholarship was not without its own unfortunate limits. The fall of the purist model was far from comprehensive. The greater part of the new writing, for example, did not involve any detailed examination or critique of scientific ideas themselves, only of their varied production and setting. Moreover, most work has been delivered in a discourse that seeks for itself some type of final truth about this most final of truth-telling enterprises. It thus continues the age-old positivist search for essences of "the scientific." The terms of this search have shifted: One now looks for how "speculation" is created as a literary strategy or how scientific norms include societal conflicts. But the entity "science" remains, as a host of activities productive of order and coherence, as a singular Academist subject. In philosophy, the case has been somewhat different (see, e.g., Fuller, 1989), but nonetheless full of attempts to compose a unified structure for the meaning and history of scientific effort. Such work, in short, from Kuhn onward, comprises a kind of late modernism of interpretation (Rouse, 1991). In this way too it reflects the larger social attitude toward "science," combining aspects of skepticism with those of worship. Even while attempting to cast new light, it has frequently ended up praising former shadows.

On the other hand, Conant's dream of healing the two cultures failed due to the institutional structure of academia itself. The scholarly carving up of the pie of "science" has helped keep such studies out of the scientific curriculum itself, certainly at any integral level. "Science studies" have developed within existing disciplines, such as sociology or history, or in new interdisciplinary fields that have kept free of any involvement in the training of scientists. This has its physical reality, since the relevant departments are almost always located in other parts of a particular campus than where the sciences are taught. The separation must be contrasted to the situation in the humanities, where the history and theory of an art form are a regular, if sometimes limited, part of each individual discipline. Thus contemporary universities continue to embody the Academist ideal with regard to "science," which remains "uncontaminated."

Finally, with few exceptions, there has been a special lack of endeavor by scholars to inform the public about the new ways of looking at science. This seems especially problematic, since scientists have been so able to argue the cause of "revolution" and therefore "service" to public goals. Again, the reasons may have much to do with academic work. No doubt, for example, one source for the lack of public interest in science studies has been a lack of interest in the public from the new breed of science scholars. Philosophers, sociologists, and historians of science, even in the midst of revealing the cultural realities and contingencies of technical knowledge, have tended to speak to each other and to no one else. The forums in which they have presented their ideas, like the language they have used, are "academic" in the extreme—removed and self-referential, interested in knowledge and in generating more scholarly work, not in what this work might mean to the "outside" world. Science studies are a possible source for profoundly changing the larger American image of science, yet one that remains loyal to a master with a shorter leash.

NEW VOICES, NEW POLITICS

In contrast to the more academic critique of the old positivism, a number of other voices did emerge during the 1970s into an oppositional politics of knowledge with regard to science. These voices have both remained relatively true to the student-centered philosophy (though not without provisos) and have also sought a broader audience, which they have achieved in attracting, to some degree, especially among educators. Their origin was in efforts to take "science" to task for its abuses of power and for its presumed epistemological prejudice against women and minorities. The belief was that technical knowledge contained a large subjective component, one oriented away from certain groups in society.

Such ideas were first put forward by a number of coopera-tives, a "loosely-related family of inquiries" (as Robert Young [1986] has called them), that, by the mid to late 1970s had begun several journals of significant stature, including *Radical Science Journal* in Britain and *Science for the People* in the United States. Other periodicals with related and overlapping interests, such as *Signs* and *Radical Teacher*, also became involved in the direct critique of "objectivity" and "neutrality." Much of this writing, at its earlier stage, was feminist (see, e.g., the collection of essays in Keller, 1985). During the 1980s it expanded con-siderably in a number of directions, including work on science education in the primary and secondary grades, where dif-ferences in gender experience were clearly indicated (Kahle, 1990). Studies of minority experience have examined both the content of scientific knowledge and the teaching methods used to impart such knowledge, and have endeavored to identify the exact ways by which "inequalities are multiplied" (see Oakes, 1985; Ballantine, 1983). This and other work, building on the critical sensibility established during the Reformist period of the late 1960s, has sought to show how elements such as textbooks, exams, lecture styles, and classroom structure all are capable of transmitting priorities, both subtle and overt, that act to exclude the values, cultural traditions, and the life experiences of other groups.[8]

Within the context of curriculum debate over science in America, these efforts represent a new kind of criticism and demanded new reforms. What they ask for, at base, is a new kind of "science" and therefore a new kind of science education. They are not Jeffersonian or Deweyan; instead they radicalize ideas of educational "equality" found in both writers. This can lead to liberating perceptions, certainly. Yet it can also devolve into a mere cataloging of injustices, a use of the textbook-as-propaganda-for-dominant-values idea as a blunt instrument of its own. Claims purporting that Greek science was entirely derived from "black Egyptians," for example, or that female scientists "see differently into nature" have proved counterproductive to an effective politics of alternative science. Such claims became more frequent during the 1980s, which brought a new episode of near-extreme conservatism in much of American education.

The more long-term and possibly successful effort, however, has been put forward by those interested in understanding scientific knowledge from the point of view of teaching it both to reveal and to reduce any subtle biases. Proponents of this method of study hope to open curricula to new styles of teaching and awareness. They face formidable obstacles, however. It is in the sciences, after all, where "facts" and "principles" have continued to dominate instruction.

THE CURRICULUM OF SCIENCE:
CHANGE AND NO CHANGE

Even during the height of the Reformist surge in the late 1960s and 1970s, the least amount of change took place in science education: Change faced its greatest resistance in this area. To introduce such things as open classrooms, role playing, or social relevance was obviously no easy task; more, it seemed ridiculous, given what science *was*, and what had to be done to learn it.

What changes were allowed to occur did so on the level of form, and could be easily dumped when the time came for a return to "basics." For a brief time students were able to choose their own experiments; their textbooks included problems adopted from real life; everyday materials were used in lab work (e.g., chemistry experiments determined the true strength of various antiacids), and so on. Beginning in the mid-1970s, after California had implemented guidelines allowing for the teaching of Creationism as a "reasonable" theory, biology texts were written with a declared "open-mindedness," so as not to claim final truth for Darwinian evolution.

With regard to textbooks more generally, there was also a general trend in various fields to focus on "ideas" and "concepts" rather than on mathematical models or finalized discoveries. This was partly influenced by Bruner and the notion he popularized that each discipline was characterized by its own "conceptual structure," a grasp of which would make all else derivative common sense. The following introductory words from a well-received and much-used text, *Physics from the Ground Up* (Carr & Weidner, 1971), indicate the flavor of this movement:

> As the title suggests, our approach to an understanding of physics is through the simple behavior of ordinary objects. We do not seek to accumulate a multitude of results, but instead to trace out a few generalizations of considerable breadth and power. Rather than covering a lot, our strategy is to uncover the most important.

This type of welcoming assurance had been utterly absent only a few years before, when the NSF-sponsored programs were in effect. At that time, students were more often introduced to a subject through thinly veiled warnings or proclamations of difficulty: "The first-year college chemistry course is now confronted with better prepared students and has an expanded and more complex body of knowledge to present" (Sienko & Plane, 1966, p. v). During the 1970s, meanwhile, textbooks that went through several editions show a marked change away from the first approach back toward this latter (earlier) type, usually with the justification that the field itself—its unpredictably rapid, "revolutionary" development—necessitates a more rigorous and demanding coverage.

Yet here too fundamental shifts never really occurred. Ideas and concepts, though perhaps less intimidating at first, nonetheless had to be memorized and applied no less than formula. The journey consisting of fixed landmarks never failed to find its final destination, either (the exam). No opening of the classroom ever took place in science beyond the earliest grades, no demand for critical thinking or individual-based instruction transpired. The latter was confined almost exclusively to the elementary school, where science became "science experiences," something akin to Dewey's notion of "nature experience," a matter of touching and observing and constructing, but wholly separate from technical knowledge and method. Indeed, much thought and effort went in to such curriculum redesign. But by the 1980s a summary of what remained from this did not hold out much reason for encouragement:

> The student-centered, activity-based curricula of the 1960s and 1970s fulfill the criteria for an ideal curriculum [but] are little used, and when they are implemented, they are frequently misused. These curricula have faded from classrooms for three primary reasons: Teachers have not understood the scientific principles the materials promulgated; classrooms have not been organized for small group interaction in science; and schools have not provided the equipment or the scheduling required. (Kahle, 1990, p. 57)

In other words, the old problem of structure that Dewey first pointed out with such historical precocity remained relevant in its depressing clarity.

At the undergraduate and especially at the graduate levels, on the other hand, research remained the ideal, all the more so by the late 1970s, given the spread of "revolution" as a professional calling card throughout the community. Indeed, by the early 1980s undergraduate students were being enlisted in more than a few universities to work on government-funded projects overseen by one or more professors. By this time too most large universities had embarked on major programs of rapid institutional growth: Thousands of new buildings and facilities, erected on the support of rising grant overhead, were going up all over the United States. Such growth reflected a new scale of competition between different schools, a race for research results that meant status and the ability to attract eminent scientists, the best students, and more federal dollars. The logic, brief and circular, had been established for more than a decade. It was backed by the enormous amount of new money being poured into DNA-related work and biomedicine generally, as well as all energy-related sciences. It was, finally, the logic of research itself.

In the end, the loss of innocence for science was merely a partial thing. For all the accusations made against it, as a source of threat or self-policed elitism, there were an equal number of claims being made in its name—claims for new knowledge, for competent children, for ecological consciousness, even for the truth about science itself. Indeed, "science" during this period was no longer the embodiment of a single national destiny; it was instead linked to many destinies, many ambitions, many possible fates. On the other hand, there seemed a radical split in the image of science. The old positivism, pronounced dead by one set of diagnosticians, was simultaneously decked in a brighter robe and given new lines to speak.

The era of the 1960s and early 1970s in public education has been termed "neoprogressive" by many writers. Yet this term is misleading. The latter-day Reformers certainly shared many of the fundamental ideals that the Progressives had popularized, and they often tried to give these ideals similar institutional forms. But with their ultimate faith in the democratic individual, these ideals hark back long before the turn of the century, back to utopians such as Richard Owen, and, of course, still further back to Jefferson and Franklin. They represent one region of the greater American landscape of political and cultural belief.

But perhaps most critical of all, the Reformist effort of the 1960s lacked the all-out devotion to "science" that was so central to Progressive concepts about society, education, and the link between the two, and that gave the movement a driving unity despite its various competing strains. No axial platform or philosophy was to be found among the latter-day Reformers. Nor, perhaps for this very reason, was there a central leader such as Dewey. The 1960s movement, after all, was based far more on criticism of what was than on optimism about what might be, and more on a desire for the wholly new than the merely improved. No strand of Progressive thought ever wanted to do away with the school system entirely. The point was to use it as a more equitably tuned instrument for adjustment. For many of the 1960s reformers, the goal was much more radical than this, on the one hand, and much more modest, on the other. Desiring an overhaul of the social order, they wanted to begin and end by humanizing the classroom experience, the relations between teacher, student, and learning. However unfortunate, these were small levers with which to overturn a system that had been in place and gathering weight for three-quarters of a century.

Troubled Symbolisms:
Science and the Curriculum
at the Century's End

By the late 1970s the Reformist movement in public education
had reeled back in the face of increasing conservative attack. In
a period of renewed uncertainty, many Americans now came to
look on its ambitions as having been misguided, and in one way
or another damaging. To historians and educators, meanwhile,
looking back from the 1980s, the brief period of attempted
reform with regard to educational equality had revealed two
essential truths. First, that massive amounts of federal support,
well organized and directed by a strong egalitarian philosophy,
could make a real difference in the lives of many children.
Second, that such an effort was by nature a fragile one. It could
be quickly and mercilessly taken apart as a result of changes in
the economy, a shift in national mood, or by special interests
eager to mainstream their own platform. Indeed, it was just at
the point when important benefits had begun to appear that
these three factors conspired toward a loss of faith in the original
vision.

Something else had been lost, something essential. While
the federal government–university partnership continued to expand, public education witnessed the dissolution of a powerful
coalition that had been building since the Progressive era
and that had peaked in the reform efforts carried out between
1957 and 1970. This coalition was made up of parents,
teachers, educators, researchers, journalists, administrators,

and legislators—power-holders for nearly every level of demand and kind of structure (Bailey, 1981). It failed to hold together because the unifying fear that had called it into being in its early years, and the liberal hope that swept it along for a few years more, died out as the country entered a period of increased confusion, distrust, and dissension.

The origins of such uncertainty were many. Inflation had severely weakened trust in the American economy. More than a few key industries had fallen on hard times in the face of increasing foreign imports. The international standing of the United States in the wake of Vietnam had fallen considerably; Americans were becoming the target of widespread criticism and even terrorism abroad. Domestic racial tensions remained high, not the least in the schools (bussing in particular became a lightening-rod issue for anger and distrust). Corporations were revealed to have polluted large areas of the country; clean up would require decades and billions of taxpayer dollars. Many ordinary foods were shown to be potentially carcinogenic, yet the medical profession was increasingly vilified as undependable, limited, and self-concerned. Faith in many institutions had eroded and political institutions in particular were increasingly viewed with distrust, even cynicism.

In such an atmosphere, Americans elected Ronald Reagan to the presidency in 1980. Thereafter, for most of the 1980s, the country swung further to the right than it had in more than half a century. Reagan promised a new era of confidence, to be delivered with grandeur and simplicity. Integral to his "conservative restoration," in fact, was the singling out of three central issues—the cold war, economic competitiveness, and liberal reforms—as a means for defining useful enemies, external and internal. All had their effect, sooner or later, on schooling. The much discussed threat of "the evil empire" led to getting tough in education to "keep America strong," a veritable repeat of 1950s rhetoric. Japan, on the other hand, became a nemesis that threatened to "overrun" the United States with foreign goods if "higher standards" and "basic skills" were not imposed immediately and in force. Finally, liberal reforms were attacked because they had supposedly injured "traditional American values" and were responsible for a weakening of intellectual fiber and will, engendered by "indulgent and atheistic schooling."

Required at all levels of education was a total return to Academist priorities (with prayer thrown in at the lower grades). "Intellectual skills" were only one part of the problem. The more pressing need was a moral one, to produce students who would serve as more disciplined, capable, and loyal "soldiers of knowledge" in the wars for dominance on the economic, military, and ideological fronts. The climate of dissatisfaction that had been building through the 1970s, backed by the general sense that America had "gone soft," was now exploited to the full. The new administration aided greatly in making the public feel good or at least justified about using the educational system as a cathartic—not as the cure, but as the disease from which all national problems began: drug abuse, crime, violence, professional incompetence, prejudice, even the national economic and technological downturn. This message was, perhaps, all the more effective given that a new baby boom of sorts had begun, especially among middle-class professionals (the original baby boomers), a fact that inevitably helped make "education" a topic of no small concern and anxiety. In the end, belief in the school as a critical origin of social reality had never wavered. In times of hope, it was itself hopeful; in times of fear, it could become a tool of threat.

SCIENCE IN THE 1980s: FEDERAL AND PUBLIC ATTITUDES

Science and the curriculum played their respective roles in the conservative restoration of the 1980s. Scientific knowledge in particular became newly popular and prestigious. As in the past, this occurred on two basic fronts simultaneously. One of these occupied the realm of public imagery, where science became the object of renewed celebration. The other front involved the federal and professional outlook, which, again, largely attempted a return to the discourse of the early post-war period. These two regions of image-making were intimately related and came to permeate a great deal of public representation, in one form or another.

Federal demands relocated the Practicalist rationale to accord with the new issues of the day. Science and technology, of course, had never stopped being promoted within government for their contribution to the economy and national defense, even during the Reformist era. But now the relationship was again made urgent, directed as it was at two particular enemies. Not only was it now essential to promote "science for national defense" and "science for economic competitiveness," it was also necessary to denote technical knowledge as a means to "repel" and "defeat" these two foreign nations, to "push back" their influences. Science was needed not merely for national honor and prestige; it was a source of raw power for America's very survival.

Under the Reagan plan money cut from nearly all social programs, including education, was transferred to various types of research, primarily defense-related. Major new projects, such as the Strategic Defense Initiative (the "Star Wars" defense system), funneled billions of new federal dollars into university, industrial, and government laboratories, stimulating inquiry in various subfields of physics, mechanical engineering, electronics, and computer technology. Simultaneous promotion of the space program, including the space shuttle, and the Voyager and Pioneer probes, repopularized the idea of "a need to keep America at the forefront of space research." Federal money and tax incentives also were used to promote innovation in specific areas of industry viewed as "under attack" by Japanese expertise, for example, telecommunications.

As a whole, the idea of "science" was firmly reattached to a variety of nationalistic goals, via concrete funding priorities. To make America again "the envy of the free world," to return it to its rightful place in the forefront of history itself, technical knowledge was an absolute necessity. Science was viewed as the ultimate source of the economic and military power the United States required to ensure its world preeminence. Science was less the engine than the ultimate fuel for America's reawakened desire to take charge of the world's future.

Integral to this overall philosophy was a quantitative argument. What America needed was not so much a better science, but a more efficient and productive one—a science able to yield more inventions, more patents, more kinds of new technology, more new industries, all in a shorter time frame. What this meant, in turn, was a need for more "technical manpower." In part the result of expansion in research opportunities, in part a reaction to federal rhetoric, by the mid-1980s many sectors of the scientific community, speaking in the name of economic competitiveness, began to proclaim a "crisis" in the supply of new scientists and engineers.[1] Also to be counted here was the image of failing American industry, adopted by many as a sign of domestic enfeeblement in certain areas of technical labor. The perceptions of a threat to the prestige of American science, of course, is one of the oldest and most sacred traditions of consensus building among scientists. Ever since the days of Dallas Bache, it has proved useful to link professional pride to foreign policy, to promote the idea of an "American science" that would further America's political destiny. American science has always managed to make America's political and economic rivals into scientific rivals; from the early days of the Republic, spokesmen for American science have successively targeted the country's major historical enemies or competitors: Great Britain, Germany, the Soviet Union, and Japan.

Journals written both for and about the scientific community therefore began to be filled with proclamations of a coming shortage of personnel. This shortage became the subject of innumerable predictions and diagnoses: "More and more," it was said, "students are not taking science and not taking math." The "education deficit," many felt, had the power to "undermine" any efforts made toward reestablishing competitiveness or a superior military defense. A standard question of the time asked: "Will we have the human capital necessary to man the system, even if we could design it right?" (Gomory & Shapiro, 1988, p. 40). The school-as-factory thus made its return as well. Policy makers, consultants, corporate advisors, and educators all believed it to be an essential insight that "a very small set of high schools . . . has produced most of its scientists and engineers" (Gomory & Shapiro, 1988, p. 41). Doubling or even tripling the number of such schools was something the country could easily "afford to do." As in the past, that is, schools and their curricula were being viewed as manufacturers of technical expertise, mills to be directed toward the war effort against foreign competition. Schooling became a large-scale concern that once again crushed away the individual, or rather that transformed he or she into raw material for production and exchange on the market of national power-making.

The fear of a coming "manpower shortage" continued through the 1980s, but then died in the recession of the early 1990s. By that time it had become evident that no such "deficit" was in the offing, and that a number of fields might even soon suffer from the opposite problem. Technical employment, in fact, had increased during the 1980s, which proved a boon to defense-related industries and to the computer industry as a whole. With the coming of the recession, rapid cutbacks took place, and the demand for scientific personnel declined.

But until this happened, and even afterward to some degree, concerns about science education continued to grow. Aided by indications of "poor" student performance on international achievement tests, and by President Reagan's demand to make American schools "first in the world in science and mathematics" by the year 2000, scientists pushed for more funding from the NSF to improve science education. Such money was soon flowing out of Washington at a spectacular rate: From a postwar low in 1982–1983 (a mere 2% of its total budget), money earmarked by the NSF for education went up without a break every successive year, reaching a requested level of $463 million (20% of the NSF budget) in 1991. The kinds of things the NSF funded mimicked those of the post-Sputnik era, at least on the surface. They included developing new curricula, supporting teacher training, and stiffening standards and qualifications, all in the interest of "producing more scientific resources."[2] Lacking to date, however, has been the intense coordination that characterized the earlier program.

In the public mind, meanwhile, science regained a measure of the heroism it had lost prior to 1980. Not only was it again tied directly to economic and military advantage, but it was constantly associated with many "positive" ongoing "revolutions"—in computer technology, in medical research, in biotechnology, and in cosmology. Again it was promoted as the source of "profound new changes" that would soon permeate everyday life, or else as the source of marvels and miracles that "defied the imagination." Indeed, the coming of the "computer age" was hailed as nothing less than a new stage in the history of Western civilization, one no less profound than the Industrial Revolution of the 19th century (much of the rhetoric of the 1840s and 1850s, in fact, found its direct descendent in the prophetic language of the "computer sublime").

Most of the new attention leveled at science tended to emphasize benefit and wonder over risk and danger. Beginning in the late 1970s, for example, technology was ritually exalted in the movies, as an origin of special effects and complex powers. At the same time, a host of science programs were aired on television, many of them aimed at informing and educating children as well as adults about "basic principles" and "the latest advances." Science writing became much more important, with most books now being written by scientists themselves and more than a few reaching mass audiences inconceivable a decade earlier. The media helped to create "science stars," such as Carl Sagan and Isaac Asimov, who gained great favor in the public eye by simultaneously speaking out against such things as nuclear weaponry and for a "new age" in all areas of scientific discovery. Indeed, in the 1980s the scientific community as a whole became a skilled advocate for its own elevation, capable in both public relations and in managing access to its inner politics.

Science journalism also underwent great expansion during the 1980s, as well as a change in attitude. Investigative reports into the possible detrimental effects of new technologies gave way to feature articles celebrating "recent breakthroughs" or "new theories." Nelkin (1987) wrote of the decade as a whole:

> Science writing, like other areas of journalism, has returned to a less critical and more promotional style—except in the face of disasters such as Bhopal or the Challenger accident. The images of heroes, breakthroughs, and frontiers that once again fill the news reflect the political rhetoric of the 1980s as well as the conventions that are the heritage of the science writing profession. (p. 97)

The overwhelming burden of all this popularizing activity fell on attempts to make science an object of interest by means of fascination. In the 1930s, "science" had been "sold" to the public by promoting its practical benefits. Now the point of comprehending technical knowledge was to come in contact with some of the "wonders" of the age. It was also, of course, to understand something about the "laws" and "principles" on which this age had been built, which must be counted as a didactic end. But when looked at closely this apparent didacticism was merely a facade for more spiritual pleasures, those, for example, to be derived from "unlocking the mysteries of everyday physics" or "understanding the forces that shape our planet." A certain quality of the scientific sublime therefore was also retained for public consumption.

The new image of science as hero did not go unchallenged. Major criticism in a few absorbant areas was by now a ritual part of the mainstream sensibility. Opposition to nuclear power and weapons research remained strong, as did skepticism regarding the medical establishment, exemplified by the alternative health movement. Antiabortionists contested any type of foetal research, while animal rights groups fought to prevent all experimentation involving animals. Biotechnology raised many eyebrows by becoming a new area of industrial entrepreneurship, a still more immediate indication of how the political economy of science was intrinsic to late-stage capitalism. Yet, in the end, such areas of conflict did little to hamper the return of broader fascination with science. Like the disasters mentioned by Nelkin, they were flares that tended to draw and contain attention around issues of immediate heat. But the general image of science was so strong, so positive, that most Americans continued to view it as more hero than villain.

Profound agreement therefore existed between the favoritism granted science by the federal government and the hope and spectacle placed in it by much of the new public interest. No doubt federal attention to scientific matters, and calls for more technical manpower were accepted as proper by most Americans. "Science" was once more employed as a diverse content for advancing the idea of "America." This was a cold war America, to be sure, which would defeat all threats to its military-industrial supremacy; yet the flip side of this, where such power was inverted into more hopeful (but no less overwhelming) images, was an America of special effects, of new wonders and blessings that were even then revealing a future full of "unimagined" possibilities.

THE MORAL CURRICULUM: RETURN OF THE OLD ACADEMISM

Given the goals of the new Republican administration, one might have expected that great priority would have been placed on science education above all other areas of training. Yet, outside the NSF, this did not happen. Nor was it a central theme in the new round of policy reports, sponsored by various foundations, that began to appear in the mid-1980s. Though widely favored in terms of funding priorities and popular image-making, science was not a priority of the federal government in terms of curriculum reform in the schools and universities.

Certainly, the curriculum attracted some attention and was treated as a political content. Many people both inside and outside the Reagan administration raised their voices to criticize what they perceived to be failure on every side. Indeed the language used was more aggressive than it had ever been: "If an unfriendly power had attempted to impose on America the mediocre educational performance that exists today, we might well have viewed it as an act of war. . . . We have, in effect, been committing an act of unthinking, unilateral educational disarmament" (National Commission on Excellence in Education, 1983, p. 2). Such were the terms set by the first and most influential of the new reports, A Nation at Risk (1983), which fixed the tone for much of the decade. Predictably, the education-as-intellectual-munitions sentiment was tied to a diagnosis based on test results: declining SAT scores since 1963 (the year the NSF-produced curricula went into effect) and poor performance (not placing first or second) on international achievement exams. In brief, American students were not taking tests very well, thus "the educational foundations of our society are . . . being eroded by a rising tide of mediocrity that threatens our very future as a Nation and a people" (p. 5).

The issue was thus not merely a loss of knowledge-as-power but instead a loss of knowledge as the source for U.S. moral superiority. This report, like others of its kind soon to follow, never addressed itself to methods of teaching, school conditions, or even materials. It offered no real plan for reform or even important change. Its main recommendation, like that of Conant, 30 years before, was simply a return to the Committee of Ten's curriculum: 3 years of English, 3 years of mathematics, 3 years of science, and 3 years of social studies, plus a half-year of computer science (notice that foreign language study is ignored). That this was almost exactly the course of study already existing in most U.S. high schools for college-bound tracks did not matter. The point was that it now should be extended to all students and, more importantly, given in heavier doses: more tests, more homework, a longer school day, a longer school year, and so on. In other words, what was lacking was not so much substance as mental discipline. Schools were "undirected"; the curriculum was "too diverse." Rather than encouraging various experiences and possibilities, education was a force that should be used to unify American culture, to impose a single vision of what it meant to be an educated individual. The quantitative argument was used here too: Quality of the learning experience is really less important than simply doing more work, in a stricter environment. The mind had to be "trained and exercised" by "constant example." The psychology of faculties was indeed back.

Between 1983 and 1989 a host of other documents appeared to harp on the same complaints for higher learning. The *cri du coeur* of the entire movement, as many historians have pointed out, was that the popularizing of higher education had rendered it demoralized and degenerate. What was needed now, if America was to be saved, was a return to an integrated "liberal arts" core, of the type that existed before the rise of the research university, which had ushered in an age of too much specialization and vocationalism. The old Academism, founded on the humanities, would rescue the nation through its "formative power." Like earlier Academist spokesmen such as Charles Eliot, Thomas Harris, and, Robert Hutchins, the new (old) Academists believed that all higher learning had an essence to it, a transcendental, even "objective" character that had to be obeyed if true education were to occur. As such, it was society that must conform to this "education"—never, to any degree, the other way around. And this involved returning to a type of learning expressive of the period that predated the important diversifications of American culture.

Curriculum reform was thus brought forward under a moral and political imperative that foregrounded the humanities. It was these subjects, with their unifying "wisdom," that would create a more coherent, shared culture in an age when "tribalization" threatened to tear apart the very fabric of the social order. To be fair, matters of instruction were also at issue. Reports such as *Integrity in the College Curriculum* (Association of American Colleges, 1985) and *College: The Undergraduate Experience in America* (Ernest Boyer, for the Carnegie Corporation, 1987) pointed to the dominant emphasis on research and how this relegated teaching—the "true function of the university"—to a low-status activity, often performed unimaginatively and as a part-time duty. Successful researchers were often rewarded— with promotions, money, perks of all kinds—over and beyond professors known to be excellent teachers, interested in the lives and careers of their students. This was something that most academics had long come to recognize and admit, at least in confidence. Writers and cultural critics such as Paul Goodman, meanwhile, had made it a topic of open (if temporary) discussion as early as the 1950s. Yet, again, no cure was offered. Where those such as Goodman began from the premise of positive change and had proposed ideas for reorganizing priorities and given departmental structures, the reports of the 1980s proved themselves critics in tone and attitude more than substance. Caught up in the spirit of the time, they began with negatives and proceeded no further, being more interested in demonizing the past than demonstrating hope for the future. Their main intent, it would seem, was to indict the university for its modernism and to suggest that any and all problems entailed in the structure of college training be miraculously corrected by a premodern curriculum.

The great attention given reports like *A Nation at Risk* encouraged various academics to speak out on behalf of similar attitudes, in an atmosphere of gaining receptivity. Alan Bloom's *The Closing of the American Mind* (1985), with its blatant contempt for students and American culture generally and its call for a return to a pre-20th-century canon of Western masterpieces, was only the most popular and popularized among a number of books that helped further the official Academist doctrine of the federal government. Bloom's book, in fact, helped set off a heated conflict within America's larger intellectual community, one that continued full steam into the 1990s. This debate pitted those who believed in the old Academism against a new breed of scholars intent on examining the sociopolitical, gender, and ethnocentric contents of literary works—a direct parallel to "science studies" scholarship (see Chapter 9) that indicates the broad-scale influence of recent European thought in the U.S. academy. Academics in this latter group, one might note, are often no less prone to "theory" and "method" and the claims of "research" than are their counterparts in the sociology of science. They are fully contemporary in having been influenced by such science-derived notions, which provide a means of generating new work for scholars and for expanding literature departments. But for these reasons too, they are viewed as "radical" by the new half-orthodoxy of Academism. They are accused of "politicizing the classroom" by those who see the curriculum as a weapon for building a more unified, moral, and communal "America."

The antimodernism of the new conservative movement must be understood, at least in part, as aimed against "science." Incarnated in the research university, in the prominence of research itself, "science" had inevitably contributed to the downfall of the classics as the ethical core of all higher learning. The sciences were important for raw power, but in the ideal curriculum they must be kept secondary. As the authors of *50 Hours: A Core Curriculum for College Students* (National Endowment for the Humanities, 1989) put it:

> A required course of studies—a core of learning—can ensure that students have opportunities to know the literature, philosophy, institutions, and art of our own and other cultures. A core of learning

can also encourage understanding of mathematics and science, and
[this book] includes these fields of inquiry. (p. 11)

Mathematics and science are therefore placed in the category of
"also," a position that immediately recalls the epistemological
and spatial meaning of the adjunct scientific schools 130 years
ago. Even mathematics, long an integral part of the liberal arts
since medieval times, is here made secondary. These subjects are
offered on the same narrative plane, in effect, as "other cultures," implying that math and science are tangential to the
mission at hand, which is said to be "the task of . . . bring[ing]
needed order and coherence."

The proposed curriculum, meanwhile, makes this still more
evident. In decreasing order of required study, one finds: cultures
and civilizations (18 hours); foreign languages (12 hours);
mathematics (8 hours); and science (8 hours). As a rationale for
this hierarchy of studies, the NEH report says:

> History, literature, philosophy, and art are at the heart of this
> curriculum because life lived in their company is richer and fuller
> than life spent in their absence. . . . These subjects have a moral
> dimension, posing questions about virtue, truth, and beauty. . . .
> Through example they engage our feelings and vivify our dilemmas.
> Through example they bring home to the heart what it means to fail,
> to endure, and to overcome. (pp. 21–22)

When it comes to sciences, the tone shifts:

> Scientists today are probing the stars and describing the human
> genome, explaining the universe and life in ways that previous
> generations never imagined. From microprocessors to CAT scans,
> technologies growing out of science affect our daily existence as never
> before. (p. 43)

One thus finds repeated the old classicist view. This, one might
say, is the anti-Jeffersonian position. Everything that the
Enlightenment had associated with "science" is here stripped
away and placed in the hands of the humanities. The sciences,
in short, lack any "moral dimension." They are about knowledge
and material power and have little to do with the "human
condition," with emotions, aesthetics, higher ideals of any kind.

Indeed, the ideal college training presented here is quite literally medieval, going back to a pre-Enlightenment split between "values"—attributed to the humanities only—and "facts," the province of science.

More fundamentally, the NEH position is that the sciences be rejected as the more important force in constructing modern society. The conviction, as earlier expressed by Conant, is that "the literary tradition . . . still underlies even the culture of the United States" (1957, p. xv). The key word, of course, is "culture." Modern life is presumably built upon ideals and philosophies that have emerged out of the Western heritage; institutions are their working expression and "culture" is their repository. Any knowledge that is separate from them is either minor or mercenary to their ultimate guidance. Literature, art, philosophy—these have provided the intellectual substance for all human societies; these are the truer embodiment of the national destiny. And thus it is to these disciplines that education must return.

How, then, should the sciences be taught? The NEH plan is taken directly from Conant once again, who argued for an "approach needed to enable the scientific tradition to take its place alongside the literary tradition" (1957, p. xviii). This approach seemed evident in the 1950s era of rising Academism, and it seemed no less so to the NEH:

> To bring science alive, to capture the moment of scientific discovery and demonstration . . . students [might] read the original words of great scientists, to see how they themselves presented their evidence, arguments, and conclusions to their contemporaries . . . [and to] encounter some of the world's greatest minds at moments of intense creativity, controversy, and excitement. (NEH, 1989, p. 45)

Science is therefore to be taught like one of the humanities, in terms of canonical texts. Indeed, unless brought into the fold of "culture" in this way, science is, finally, "dead"; only the model of literary study will suffice to ressurect if from the glass coffin of its own knowledge. If the national destiny is, or should be, a recovery of "needed order and coherence," science and technology can be made to participate, but only when transformed into literature.

Such thinking therefore formed a clear reaction against both the popularization of higher learning and the more basic reality of the scientific foundations of modern society, the pervasiveness of science and technology in everyday life. These factors had continued to leave the humanities in a confused historical position, without a clear and stable purpose beyond themselves. Claiming the moral high ground was a discursive trope, certainly. But now, in an era of political conservatism and external "threat," this too could be made more urgent, connected as it was with the fate of the nation itself.

MAJOR TRENDS AND INFLUENCES IN SCIENCE EDUCATION

Platforms such as those offered by the NEH and other like groups were not the only ones to affect science education during the 1980s. Indeed, as a consequence of the large amounts of money being funneled through NSF programs, science education in general became a field of considerable academic study, most of all in the primary and secondary grades. Patterns of argument for reform have been varied, and have been brought forward by scientists, educators, psychologists, and others. Yet when examined in the light of history, recent trends of thought can be seen to repeat familiar patterns. More than this too, they can again be shown to have depended on "scientific" ideas no less than in the past.

As mentioned, for example, the most recurrent "proof" of the need for change was declining test scores. Here the comments of scientists and many educators were in complete agreement with the new Academists. One recent report, *Science for All Americans*, sponsored by the American Association for the Advancement of Science, offers typical criticism:

> One only has to look at the international studies of educational performance to see that U.S. students rank near the bottom in science and mathematics—hardly what one would expect if the schools were doing their job well. . . . The United States should be able to do better. . . . [It] has deliberately staked its future well-being on its competence—even leadership—in science and technology.

> Surely it is reasonable, therefore, to expect this commmitment to show up in the form of a modern, well-supported school system staffed by highly qualified teachers and administrators. (Rutherford & Ahlgren, 1990, p. vii)

Scores on standardized tests are invoked as the revelation of two sad "truths": first, the entire nation is failing in leadership; second, that nearly *everything* is wrong with the American school system. The basis for such a remarkable leap of supposition is, however, never given, let alone subjected to inquiry—and this in a book that argues the great need for "critical thinking" while impugning "the learnings of answers more than the exploration of questions."

Faith in test performance as a kind of scientific indicator for widespread conditions, like faith in the "scientific attitude" as a solution to all human problems, represents the applied positivism that has surrounded so much of American schooling in this century. Indeed, as a basis for critique, test performance becomes itself a technical argument for exactly the type of education that has been universally decried for over a generation, that is, education-as-test-preparation, as memorization, as recitation, as a mechnical process in general.

But a reliance on testing was not the only response to appear in pedagogic circles during the 1980s. Another was linked to recent developments in cognitive psychology, particularly to themes that emphasized the mind as a "processor of information" and the teacher as a controlling force with regard to decision-making tactics that guided the "flow" of such information in the classroom. The basic model was the computer; "intellectual skills" were largely correlated with "learning strategies," and the individual mind, not the external environment (as in behaviorism), was posed as the final determinant of what was absorbed and how. Such ideas became widely popular by the late 1980s, and appeared in a number of policy reports, such as *Tomorrow's Teachers* (Holmes Group, 1986) written by the Holmes Group, a task force comprising representatives from over ninety major universities.

This report, like others in its class, ended by playing several games at once. While stating that learning is student-centered by nature, it nonetheless held up the teacher as the key agent in the classroom. It also maintained that "competent teachers . . . possess broad and deep understanding of children . . . [and] combine tough-minded instruction with a penchant for inquiry" (p. 3). Good teachers, not students (as for Spencer and Dewey), were scientists of a sort. Classrooms were indeed "laboratories," but the importance of "information flow" meant that learning had to be "managed" and "directed." The overall image therefore embodied ideas not of exploration but of corporate organization, "big science" in a word. Teachers failed in the past not only because they did not know enough about their subject, but also because they did not understand the technical and organizational aspects of learning and had therefore been unable to embody them. In the end, the relevant pedagogy therefore obeyed two widely advertised goals during the 1980s: the call to advance "intellectual skills" for the sake of international competitiveness and the demand for giving new status (and responsibility) to teachers. Together, whatever might be said about student-centered learning, these goals only shifted the philosophy underlying the given system of schooling. "Education" was still mainly about the production of competence.

For other educators, meanwhile, the combination of test scores and "poor" teaching could be read as support for a long-standing suspicion. Namely, that the coming "manpower crisis" was more a problem of decaying interest than anything else. The reasons for this problem and what could be done about it seemed clear to many:

> Students don't see the relevance of science to their lives. Most of them reported having had only limited opportunity to engage in active and meaningful learning [and to make] connections between what they are taught in classrooms and events in their own lives. . . . Alternative teaching methods, patterned after the methods of science itself, may provide opportunities for more meaningful learning. . . . Classrooms engaged in such a "spirit of science" approach would add many of the investigative procedures that scientists use, such as observing, measuring, and hypothesis testing. (Jenkins & MacDonald, 1989, pp. 62–63)

Historically speaking, this proposal represents a mix of diagnostic ideas from the 1960s and the NSF-sponsored pedagogic therapies of the late 1950s. Yet the image of what scientists do, their "investigative procedures," seems even older than the NSF program—indeed, it is largely the kind of old-fashioned conception that scientists at that early time felt had to be jettisoned.

It would be wrong, however, to merely dismiss this proposal as simple recurrence. Rather, it seems to indicate that older language continues on in one form or another as part of the larger institutional memory and discourse of the educational field as a whole. Fixed by a particular epoch, the terms and ideas attached to "reform" are added to the larger rhetoric and therefore tend to resurface when certain questions are again asked.

Something of this same kind of historical mixing, for instance, can be found in most other proposals made over the past 12 years regarding the purposes of technical education. The contemporary importance granted to test scores does not prevent a work like *Science for All Americans* (Rutherford & Ahlgren, 1990) from employing terms reminiscent of "the scientific attitude" of the 1930s: "Scientific habits of mind can help people in every walk of life to deal sensibly with problems that often involve evidence . . . logical arguments, and uncertainty" (p. vi). Another recent document, yet more striking in its combination of past rhetorical strategies, is that entitled *The Content Core: A Guide for Curriculum Designers* (1992), published by the National Science Teachers Association, which can be read almost as a catalogue of criticisms and proposals that have appeared since the turn of the century:

> Clearly, the time has come for the educational edifice of facts and the layer cake to be dismantled. (p. 14)
>
> Students also must see how science directly relates to their lives and larger human concerns. (p. 15)
>
> Science programs should involve appropriate sequencing of instruction, taking into account how students learn. (p. 15)
>
> A course organized around a Great Idea (or Ideas) of Science necessarily integrates those disciplines that operate under its laws or principles. (p. 19)
>
> An inquiry-based science program requires performance-base assessment. . . . Outcome statements are made operational by the kinds of problems and tasks students can perform at different times. (p. 29)

Represented here are all the major movements in 20th-century educational philosophy: "Relevance," child study, Progressivism, neo-Academism, scientific curriculum-building. And indeed, the research on which this document was based comes from a variety of fields, including psychology, ethnic/gender studies, history of science, philosophy of science, and, of course, pedagogy too. Each of these areas has taken its own view on science education, asked its own questions, and therefore has tended to adopt a given discursive pattern. Developmental psychologists return to the language of child study; behaviorists use the jargon-filled discourse of "assessment"; historians of science point up the datedness of the scientific model; pedagogs speak of "objectives" and "appropriate learning sequences." Each of these interests can overlap with others; yet taken together, they provide not a comprehensive vision of what is and might be done, but instead a stew of more-or-less competing ideas.

The result, finally, is a lesson in itself. For when one closely examines the curriculum and instructional advice of this report, it is evident that there is little new in it, little in fact that differs to any profound degree from ideas and practices that have existed for some time.[3] Indeed, the same might be said for a great many of the reports on science education that have appeared during the 1980s and early 1990s.[4] Such documents tend to fall into three basic groups that have come to dominate the field: those that advocate teaching science as inquiry (students as "little scientists"); those that emphasize "relevance," that is, teaching science in a manner that addresses actual problems in students' lives; and those that argue for teaching science as "conceptual change," according to the psychological research. Again, the mixture of historical philosophies is clear. Each group wants to help students learn how to "think scientifically," to develop "scientific literacy." Each tends to begin with a Deweyan proclamation of the need for more student-centered learning. Each bemoans the rule of textbook science. Each stresses a "hands-on" curriculum and speaks of the successes revealed by recent and prior experimental programs along these lines.

Yet each historical philosophy, as a final answer to science education, proved to have its limits (or excesses), and these too tend to be repeated today as well. When it came time, for example, to evaluate any new "experiments" along child-centered lines, researchers in each group have turned immediately once again to test scores, the behaviorist standard since the days of Thorndike and Terman. Moreover, the burden for nearly all change sooner or later falls on the teacher, whether this involves ensuring "appropriate-level activities," creating a "learning community" in the classroom, or taking extra care to counter the socializations favoring boys over girls in science. As a result, the more detailed nature of recommended reform is only too often a matter of fine-tuning the existing system, for example, writing better texts, connecting scientific concepts (which are fixed and invariant) to "real life," doing a better job of correcting student explanations, and so on.

No doubt the most radical-seeming of the "new" philosophies in science education is that known as "constructivism," which has come to serve in whole or in part as the conceptual base for all three groups mentioned above. Constructivism represents the most direct influence on pedagogic theory that has come out of science studies scholarship. It seeks to bring together an image of what science is, derived from sociology, with an image of how learning proceeds, adopted from the developmental psychologists Jean Piaget and L. S. Vygotsky. Simply put, it holds that knowledge is not a fixed type of substance, passively received and absorbed, but is instead actively constructed by each human being. What is called "scientific understanding" must be built up anew by each individual through the organization and adaptation of his or her own experience. Objectivity is a complete myth; subjectivity (though often in shared form) is the only reality. The implied model for learning is therefore said to be entirely, even absolutely, student-centered. What the teacher supplies is mere "sensory input."

During the late 1980s and early 1990s constructivism became a unifying concept for much work on science education, a kind of rallying idea for rethinking teaching and the curriculum with the goal of putting the student first. Though not a philosophy self-knighted with the "scientific," it is involved with various psychological theories of learning that do make such claims. Thus, while it seeks to represent itself as the polar opposite to the old positivist view, its followers within the field of pedagogy are no less prone to view it as a means to better promote such things as "higher order thinking," "process skills," and "conceptual change." And even with regard to its "Robinson Crusoe model" of learning, educational constructivism can not overcome problems that have existed for a long time. For example, not only does it ignore the issue of knowledge as a form of social conformity, and promote the old mass efficiency argument via testing and "skills," it posits the teacher as a type of beneficial manipulator of possible "inputs," and education itself as a matter of "making sense" of given "data." It is thus more or less a subjectivized computer model, presented as an objective situation for learning. It is one more theory, in the tradition of G. Stanley Hall and the Herbartians, that tries to put the human dimension back into learning yet in its own fashion removes it once again.

The computer, one might note, is only the latest model to offer a kind of technological solution to questions of learning, cognition, and behavior. Its adoption into theory as an intellectual prototype almost exactly correlates with the period in which it was taken into the classroom as an intellectual tool. Indeed, one could point to three basic stages in the postwar use of "outside" technologies for school instruction. In the 1950s movies were introduced by means of teaching films as part of NSF-sponsored reforms. By the late 1960s television became an often-tried "experiment." Both attempts to "revolutionize" the classroom were promoted on the basis of making school more interesting and attractive by bringing in daily educational experiences. But at a deeper level, the intent was almost purely manipulative: Statements of theory that underlay such use spoke of "stimulating" and "motivating" students, of gaining "targeted responses"—behaviorist concepts that reconfigured the belief in mass methods for mass results. The computer now extends this line of reasoning into the field of "cognitive processes" and "outcomes-oriented education." The language has changed, but the fundamental view of the student remains the same. "Student-centered" does not necessarily mean "student-empowered." In the face of such language, Dewey's feeling that learning must inspire—not form or pattern—"the meaning one finds in the affairs of life" seems just as radical today.

Perhaps it is not surprising that all committed attempts to teach students how to think "scientifically" are confined to primary and secondary schooling. Almost nowhere, that is, does one encounter a similar program for undergraduate college education. As I have pointed out in previous chapters, this has everything to do with concepts of the student, which are tacitly divided between different age levels. John Goodlad, for example, in his 1984 study *A Place Called School* notes:

> An interesting shift in orientation made itself apparent . . . at the fourth grade level. Teachers in the first three grades frequently appeared to place the child's personal orientation to the natural world above . . . mere possession of information about science. By contrast, many teachers in the upper three elementary grades appeared simply to be teaching the children about some of the topics and methods of science. (p. 213)

The actual practice of student-centered learning therefore seems confined to ages eight and under. In this age group the lessons of "growth," the sentiments of child study, the very idea of the child, find their historical legacy. Beyond this group, the student enters the realm of rated performance. He or she is now a mature object of mass pedagogic treatment. It is no coincidence that this is precisely the age where the textbook and the lecture become the standard learning tools (prior to this, a range of materials, situations, and group learning are employed). What, then, of the upper end? A different type of citation, echoing an earlier generation of intent (e.g., James Bryant Conant), states the case: "Secondary school science courses in the United States are major filters to careers in science and science-related fields" (Aldridge, 1992, p. 1). Once one enters college, learning is transformed into training; the student crosses another threshold, this time into the ranks of novitiate.

This is why trends of policy for higher learning have often been unrelated to or even opposed to those in the lower grades. Only during the late 1960s was any effort made to bridge this traditional gulf. In the 1980s, meanwhile, it was everywhere in evidence again. For instance, reforming study materials for undergraduates was never really a topic of discussion, even though textbooks and facts still predominate at this level too (something that should perhaps give one pause for thought). On the contrary, instead of making upper-level learning more "relevant" or inquiry-based, many within the technical community have taken up the government's own Academist position and argued that the decline in U.S. competitiveness reflects a loss of "leadership," and that the way to restore this is to require scientists and engineers to "pursue a liberal arts program during their first 2 years of college, reserving the study of engineering [or high-level science] for their last two—but even then, without departmental specialization" (Hutchinson & Muller, 1988, p. 71).

In the end, like the policies for changing secondary science education, programs aimed at undergraduate learning often cancel each other out. Between educators who are worried about student interest, those who want future teachers to be trained "in depth," and those who seek wisdom in a "liberal arts" core, the system that exists will undoubtedly find no long-term reason for important change. Indeed, one of the most significant trends in real practice has been the extension of research-apprenticeship downward, into the senior and junior years. Begun in some of the major universities at least as early as the mid-1970s, this trend began to influence smaller liberal arts colleges, even those without graduate departments, in the 1980s. Students perceived as either "gifted" or "highly motivated" have been increasingly recruited to help perform graduate-level work on major grant projects, especially in areas of heavy funding. This means that it has become more possible for science majors to become coauthors on published articles (something still inordinately rare in the humanities), to therefore be in some sense *real* scientists— not merely simulated ones—even before reaching the final rung on the training ladder. The system functions to help select and encourage a chosen group, to accelerate their novitiate. It is a means to professionalize the undergraduate experience, a goal that stands immediately opposed to the humanities Academist ideal yet entirely supportive of Academism in science. But its final implication should not be avoided. "Scientific thinking" is not to be separated from "scientific acting," and this everyday practice tends to mean "big science" most of all—cooperative, yes, but also hierarchical, obedient, competitive, and above all, profession-centered. Given this as an endpoint, much of the effort in recent science education may be more influenced by humanist sentiments than has previously been admitted.

CONCLUDING REMARKS

Between 1970 and 1980 total enrollments in U.S. higher education rose from roughly 7.5 million to nearly 12 million. By 1989, however, figures show only a moderate increase to around 12.8 million, of which nearly 40% were part-timers and an ever-increasing number were enrolled in junior and community colleges, where vocational education is more the norm. Many of the new students are adults, especially minorities and women, who are seeking education for immediate career needs. Among entering freshmen in 1989, meanwhile, the following breakdown of intended majors has been recorded: 24.5% in business-related studies; 10.2% in engineering; 9.2% in education; 8.7% in the humanities; 5.9% in the natural sciences (Oakley, 1992).

Higher learning in America is currently a system of training for employment. Whatever elevated goals might be proposed for it, the realities of this system in its greatest era of popularization make it an instrument that mirrors, participates in, and creates social realities, economic ones in particular. One is struck at how this fulfills a certain part of the Progressive program. Certainly, it is interesting that education as a career choice has more appeal than the humanities, and nearly twice as much appeal as the natural sciences. Yet to read this as a kind of final, sobering reality—especially with regard to "intellectual standards"—is to miss the point. Inasmuch as it trains for employment, college also trains for both mass and elite endeavor.

Indeed, the hierarchy might be turned around, to form the classic pyramid of social empowerment. Because they attract only a relatively small number of students, the sciences remain more dense in real-world power than any other area of knowledge. As Paul Goodman once noted, the distribution of student interests in higher learning reflects a basic top-down corporate structure. Elitism is the very function of limited appeal in powerful disciplines backed by perceptions of hard-to-learn content, by exams and qualifications that "weed out" the majority, (a garden analogy that should be examined more closely for its aesthetic and moral implications), and by "advanced" courses that bribe the self-importance of those who succeed. "I am unimpressed by the argument that the demands of mass education and expanding population make the schema inevitable," Goodman said. "In a period of historical transition . . . we should be emphasizing human scale" (1968, pp. 388–389). These comments on "the education industries" were written nearly 30 years ago. They clearly repeat the Reformist ideal that learning might one day be recognized as a personal affair—not simply a subjective one.

Are student interests created or socialized by schooling? This is not merely a cognitive question. Yet, in the end, it may be the wrong question to ask. Just like the battles fought over the literary canon, the call to "scientific literacy" assumes, first, that school fixes sensibility, and second, that it is the one and only chance students will ever get to learn about science (or to read the classics). Meanwhile, in one of the most comprehensive studies ever performed on American secondary schools, Goodlad (1984) has reported:

> Developing "the ability to read, write, and handle basic arithmetical operations," . . . pervades instruction from the first through the ninth grades and the lower tracks of courses beyond. What the schools in our sample did not appear to be doing . . . was developing all those qualities commonly listed under "intellectual development": the ability to think rationally, the ability to use, evaluate, and accumulate knowledge, a desire for further learning. Only *rarely* did we find evidence to suggest instruction likely to go much beyond mere possession of information. . . . Nor did we see activities likely to arouse students' curiosity or to involve them in seeking solutions to some problem not already laid bare by teacher or textbook. (p. 236)

As of 1990, there was little to suggest that this situation had changed for the majority of American schools.

It is thus not the creation or training of interest that might be examined but instead the reception it gets. Such instruction by its very nature singles out those who have the greatest ambition, confidence, or "talent" (which usually asserts itself at some point, in terms of performance). If, from the federal perspective, the final goal of education is to filter out future leaders—in science, let us say—such that America can remain an elite among nations, then this system is already an excellent base for doing so (see, e.g., Spring, 1976).

No doubt it is for this kind of reason, mixed with a dose of fatalism, that things have not changed very much on the most fundamental plane. Major reforms even into the 1990s have seeped in from the periphery, causing the borders of the system to expand and contract, to change color and, at times, permeability. But the most deep-seated changes have always remained on the level of language—and in much of this, too, modification has been limited. Surely this is why the same criticisms and therapies have also tended to recur with such depressing regularity. In a certain way, the tradition of American utopianism reasserts itself each time a new policy report is brought out, offering a whole-hearted prognosis for the ails of national education. Given Dewey's point about the scale and structure of this system, it can scarcely be imagined that the efforts of a few years will ever be able to infuse themselves very deeply. More time, money, and commitment are always required. Any understanding of this, however, will not halt the ceaseless demand to make things "better," nor the use of such demands to further political interests. These things too are part of "education."

Yet, in reflecting on the conclusions of those such as Goodlad, one cannot help but go back to Jefferson, with a certain humility for the period so often dubbed "the American century." The vast hopes that Jefferson placed in national education, in the teaching of science most of all, remain, certainly, but today have largely developed into the prominence given to sectarian critique, themselves indicative not of special interests alone but of the long-lasting and perhaps undying desire to somehow, finally, make things right. "Science" has given its lessons here, as the model for knowledge, as the personification of national greatness, and as the credential for systematizing every aspect of schooling—from the structure of administration to the structure of the mind. Yet these lessons have not themselves always been hopeful ones. Too often, perhaps, they have been taken to suggest that "getting things right" means finding a formulaic solution to the many questions that schooling will always pose. As such, they have often sought something quite different than "the precious blessing of liberty" for the student or the future. The great aspirations and the grand tragedy of Jefferson and Dewey must be understood to lie in their unbroken relevance to each new era of reform.

Epilogue

Jefferson and Franklin argued against the Academism of their time. It was their belief that education in the new republic would overcome its narrow, Old World allegiances mostly through an expanded curriculum in the sciences, a curriculum that would connect the student with the needs and life of the new society growing up around him. Their view was that education could advance the general welfare through individual empowerment. It would build that welfare, it would help forge the new society because it would begin with the individual, with his life and needs. Learning, Franklin once said, was a matter of living more broadly, generously. There is a line, then, that runs from this aspect of hope for America, a hope largely based on science, through that of early socialist reformers like Robert Owen, through the ideals and plans of John Dewey, and through Reformism as a whole (in its committed, nondogmatic forms) up to the present day. All along, one might say, the hope has been for an education that would be responsive to the democratic self, and to the possibility for a more open society, undergoing continual nonviolent "revolution," rather than to a vision of tightly restricted leadership and a centripetal "community." It is this argument, then—between "education" as a search for new freedoms, on the one hand, and as a desire for imposed securities and final solutions, on the other—that forms the link, stretching across two centuries, between Jefferson and the NEH.

History, however, has not sided mainly with Jefferson. Many claims made in the name of "science" have looked to the other position. The search for "scientific" answers in schooling has often been an attempt to rationalize mass education, especially in the form of techniques and methods that seek to guarantee some type of "output" and therefore make the individual essentially irrelevant. By no means has this type of thinking dominated all ideas of activity. Indeed, the tradition of local "experiments" in education has always been based, even before Dewey, on concepts of child development, learning potential, and the like—on "science," that is, as a liberating teacher. Such experiments in student-centered learning continue today throughout the country, often in the form of local, small-scale programs supported by government funds and designed by experienced educators whose intent is to remake the educational system in miniature. In the past, programs of this type have been lauded time and again for their successes. But the great majority have never lasted very long, and even where they have, their influence has rarely extended beyond their own borders. In some part, this is due to their economic vulnerability to changes in political climate. But it is also because on the other side of their fragile edges a different type of "science" tends to rule.

True reform in American education—not just the revitalization or cashing out of existing structures—has always been left in the hands of revolutionaries and experimentalists. This is its tragedy and, in some part, its waste. For over a century, the schools have taken children from the home and rescued them from it. They have taken on this power in a society where the child has precious little contact with any adult who is not a teacher. Yet, as Dewey noted, the edifice of schooling, with its large classrooms and tight scheduling, tends to prevent vital contact here as well. To drop the systems of grading, to abandon the blackboard, to give up the "science" of objectives and assessment—this would mean hurtling the teacher and the administration into a set of personal relationships with students from which there would be no return. Dewey thought a class should be no bigger than a large immediate family. It was his perception that learning was always an individual affair, requiring security and enjoyment, whereas "education" referred to institutions, both architectural and philosophical, within which learning is supposed to take place (and which include more than just the school itself).

However appealing, these are not the sort of ideas promoted by the more mainstream intersections between science and education. "Science" in this case would tell us that good teaching can be ensured by method. Such method, in turn, can ensure the production of good learning, therefore (for example) good scientists, good science, and whatever else is presumed to follow therefrom (discovery, national power, economic advancement, etc.). The Reformist argues that good teaching does not exist as such, or does not matter by itself, as method. To him, such emphasis takes the focus of attention away from the student and devotes it instead to institutional wants and results. The search, instead, is for good learning, which takes place because of a sharing of responsibility and commitment and which progresses furthest when the teacher is left behind, like an abandoned husk, or transformed into a mere guide or medium. Political connections are important in this respect. For those in the Jeffersonian, Deweyan line, who were interested in continual betterment, the final goal was what a student could do and be as a living member of society, not what he or she might achieve measured against some fixed standard of "the educated person." The Reformist view is that the aim of teaching, in fact, must be self-erasure: to bring the student, whether in the sciences or the humanities, to the point where he or she casts off all dependence and looks to self-learning as the only future. This has been expressed many times, by many different writers, in many different ways, especially where the subject of college and its social role is concerned. Paul Goodman, for example, once lamented that "the university, which should be dissident and poor, has become the Establishment. The streets are full of its monks" (1964, p. 71).

Nonetheless, the faith that the curriculum retains a power of conversion and creation, especially when guided by the teachings of "science," remains no less fervid today than a century or more ago. Whether favored by those on the left or those on the right, "research," "assessment," "performance," and "methodology" continue to dominate pedagogic thinking and to support the systems of schooling exactly as they have come to exist. The final irony may be, however, that much of this makes little difference in the end. Schooling at any level may in fact be far more neutral than the overburdened expectations that are now a part of "education" would ever allow. "We are all so committed to education that we forget this," writes William Boyd (1966),

> Yet if our history of the theories, mechanics, and administration of schooling teaches us anything, it is surely that ultimate educative value [sic] depends on the formative engagement of a person with life's challenges and opportunities outside any school. What made an 18th-century gentleman humane or scholarly was certainly not his schooling. Yet from that time on the Industrial Revolution has communicated the idea that schools may make or process gentlemen, scholars, and experts. (p. 470)

No doubt too, what will make a more productive society, economically or epistemologically, or a more equitable one, genderally and ethnically, will also come from factors that reach from beyond the doors or gates of schooling. And science, in whatever form, to whatever pedagogic aim, will itself remain endemic to such a limit.

Notes

Introduction: "The Field at Our Door"

1. All Jefferson material is quoted from *Writings* (Peterson, 1984).
2. It should be mentioned that the word *science* had a different meaning in Jefferson's day than at present. For leaders of the time, it generally meant a combination of what we think of today as scientific-type knowledge (of the material universe), empirical and natural philosophy, (theorizing about the origins and meanings of "nature," including "human nature"), technology/technical methods (including, that is, not merely devices and machines, but practical processes in almost any area of expertise, from agriculture to architecture), and, according to Samuel Johnson's *A Dictionary of the English Language* (1755), "comprehension or understanding of truth or facts by the mind." Such meanings remained in place with regard to general attitudes among the public for much of the 19th century.

Chapter 1: Science and Democracy: Emerging Trends of Faith

1. Again, Jefferson provides excellent articulation: "These [scientific] societies are always in peace, however their nations may be at war. Like the republic of letters, they form a great fraternity spreading over the whole earth, and their correspondence is never interrupted . . . Vaccination has been a late and remarkable instance of the liberal diffusion of a blessing newly discovered" (Letter to John Hollins, February 19, 1809; in Peterson, 1984, p. 1201).
2. During much of the revolutionary period colonists were subjected to British embargoes or were urged not to buy British goods. The result was

a period of distinct austerity. After the war, ironically, trade was quickly reestablished and a flood of English products entered the United States. The British, meanwhile, sought revenge against the American victory and to head off possible competition by closing many of their own markets to American goods, thus enhancing the economic imbalance through severe trade restriction.

3. See Kasson (1976, pp. 36–40). In this connection, it is also important to point out the clear preoccupation that colonial leaders of every stripe had with the idea of power, its origin, meaning, and control. Power was viewed as a potential evil, a temptation toward self-aggrandizement, if not legitimately guided by forms of mutual consent. The images used to describe it were almost wholly anthropomorphic or biologic: it "trespassed," "encroached"; it resembled devouring "jaws," a "cancer," a "seizing hand" (see the excellent discussion in Bernard Bailyn's classic study *The Ideological Origins of the American Revolution* (1967, pp. 55–63). Such organic images—fixed on a kind of unleashed and savage vitality—were then directly countered by a civilizing political imagery of restraint and control attributed to science was a "republic" dictated by "reason," a "fraternity," or a new kind of "community" governed by universal rules of demonstration and allegiance to shared goals.

4. The phrase is Jefferson's; see his *Notes on the State of Virginia* (1787).

5. Much of this sort of thing can be found, for example, among the writings of visiting or recently settled European geologists, in whose work the direct and literal association between certain republican values, brought to a kind of culmination, and the American landscape itself can often be seen, sometimes in fascinating ways.

6. Here, one might note certain long-term connections. This type of study, focussing on a fixed series of sacred texts, was prevalent throughout the Western world during this period, and was itself a direct outgrowth and continuation of the curricular standards created during the later middle ages. These standards, in turn, were themselves derived from an earlier monastic tradition, merging exegesis of the Bible with that of a selective grouping of canonical classical *auctores*. The basis for this tradition, meanwhile, was the Christianization of classical learning itself, whose purpose was the pursuit of virtue and whose teaching, during the declining years of the Roman Empire, was divided into the so-called "liberal arts." As defined within the medieval university, these "arts," in order of hierarchical importance, were the following: grammar, rhetoric, and dialectic (known as the *trivium*, or "three ways"), and arithmetic, geometry, music, and astronomy (the *quadrivium*).

The origin of the word "liberal" with regard to these disciplines remains obscure. Nonetheless, it came to have a specific meaning for medieval authors and teachers. As noted by Hugh of St. Victor (early 12th century), for example, study of these "arts" required a mind that was "liberated" from worldly concerns; moreover, "only free and noble men were accustomed to study them, while the populace and the sons of men not free sought operative skill in things mechanical" (in Brown, 1975, p. 64). A more blunt way of putting this is offered by Ernst Robert Curtius, who states that the "liberal arts" were "studies whose purpose is not to make money. They are called 'liberal' because they are worthy of a free man" (1953, p. 37). Freedom, in short, was conceived within this tradition as a matter of removal and elevation from common problems of life, something that, during the late Roman period, was connected to an understanding of ubiquitous corruption and degeneration and that later was attached to belief in the ultimate importance and virtue of contemplation. The original scheme for the "liberal arts" seems to have been set down by Varro in the 3rd century B.C., at which point architecture and medicine were also included. These disciplines, however, were later felt to be overly linked to "operative skill in things mechanical" and were dropped. The seven that remained were thus the "purest" of all. Though their end purpose may well have been reflection of a higher type, they nonetheless came to serve magnificently well a new era of empire building and expansion, with its greatly increased court and its need for a restricted elite of courtiers, diplomats, loyal aristocrats, and others who commanded the powers of language and persuasion.

It should be noted, finally, that "science" remained toward the bottom of the "liberal arts" curriculum, and was, in any case, dominated by classical mathematics. Indeed, historians have often pointed out that the idea of science did not yet exist as such but had to develop outside the university system altogether. Francis Bacon, of course, is the prime example here; but the majority of his followers in the Royal Society of London were also of non-academic professions. Powerful as their influence was, the ideas they framed were taken into the universities according to existing categories: Science became part of the Arts curriculum and was steeped in its teaching with a mixture of philosophy, rhetoric, classical languages, and so on. Anything vocational—anything of the "manual arts"—was strictly omitted. The sense that "the mechanical arts" were "illiberal" never waned and reflects the long-term conservatism of university learning, which from the very beginning preferred to heroify itself through the image of the elite "free man," always in opposition to those "not free" who, in the end, were tainted by everyday concerns—those, to be sure, that provided the material base for the "liberal" mind to exist.

7. Indeed, as Nietzsche (1877) suggested, nearly the entire foundation of knowledge, and for that matter political power, was often still based during this time on modes of persuasion, of constructing and imposing territories of belief, obedience, and truth. For a more scholastic treatment of this point, see also Morrison (1983).

Chapter 2. Science in the New Republic: Faculty, Family, and the Failure of Idealism

1. With regard to the absence of patronage, one noteworthy exception was provided by Benjamin Thompson (Count Rumford), who made at least two important endowments during these years: the first, in 1796, a gift of $5000, to be set up in trust to the newly established American Academy of Arts and Sciences for the sake of research into the nature of heat and light; the second, in 1815, a willed fund of $1000 to Harvard College for establishing a course of lectures aimed at teaching "the utility of the physical and mathematical sciences for the improvement of the useful arts, and for the extension of the industry, prosperity, happiness, and well-being of society" (in Noble, 1977, p. 21). The Harvard chair was filled by a physician (Jacob Bigelow, popularizer of the term "technology"), the closest thing to a professional scientist at the time.

2. A humorous yet telling anecdote concerning a classroom science demonstration at this time—indicating the image of the "science professor," the role of the student as passive observer, and the type of equipment that had to be relied upon for such "experiments"—is offered by Washington Irving in the first few pages of his *A History of New York, From the Beginning of the World to the End of the Dutch Dynasty* (1809):

> In the course of one of his lectures, the learned professor [Von Puddinghead], seizing a bucket of water swung it round his head at arms length; the impulse with which he threw the vessel from him, being a centrifugal force, the retention of his arm operating as a centripetal power, and the bucket, which was a substitute for the earth, describing a circular orbit round about the globular head and ruby visage of [the professor], which formed no bad representation of the sun. All of these particulars were duly explained to the class of gaping students around him. . . . An unlucky stripling, one of those vagrant geniuses . . . desirous of ascertaining the correctness of the experiment, suddenly arrested the arm of the professor, just at the moment that the bucket was in its zenith, which immediately descended with astonishing precision, upon the philosophic head of the instructor. . . . A hollow sound, and a red-hot hiss attended the contact, but the theory was in the amplest manner illustrated . . . whereby the students were marvellously edified, and departed considerably wiser than before.

3. See, for example, the essays dealing with the early 19th century in Abir-Am and D. Outram (1987).

Chapter 3. The Age of Jackson and After, Part 1: Popular Imagery and the Idea of Science

1. Later in the century lyceums were effectively brought back in the form of the well-known Chatauquas, or more precisely the Chatauqua

Literary and Scientific Circle, created in 1878 by John Heyl Vincent as "a school at home . . . a "college" for one's own house" (Vincent, 1886, p. 73). In its original form, this was essentially a mail-order college curriculum, to be studied over four years either by individuals or in a discussion group, with diplomas granted at graduation. Like the lyceum, it was often the only cultural organization outside the church in rural areas, and its overall popularity was no less striking: From an enrollment of about 8,000 the first year, it grew to over 300,000 by 1918, during the Progressive era. Its curriculum was based mostly on reading, on discussion of texts; but observational science (astronomy) was also pursued. The Chatauqua movement was much less dependent on the lecture than was the lyceum, and thus held more closely to the ideal of self-education. Yet its model was clearly the "great books" college. More, its popularity had a great deal to do with three things: the great importance placed on education in the Progressive period; the continuing status of the "cultured individual"; and, most practically of all, the intimidating character of institutional higher learning, which, by the 1890s, had become highly professionalized as a result of the rise of the new research universities and the imposition of strict entrance standards in the older colleges (see Chapter 5).

2. The passage from "The American Scholar" (see Emerson, 1983) that is most often cited in this connection is surely the following:

> Every thing that tends to insulate the individual,—to surround him with barriers of natural respect, so that each man shall feel the world is his, and man shall treat with man as a sovereign state with a sovereign state;—tends to true union as well as greatness. . . . The scholar is that man who must take up into himself all the ability of the time, all the contributions of the past, all the hopes of the future. He must be an university of knowledges. (p. 70)

3. One should recall that the tenor of the age was often high-strung and inspirational, and listeners seldom demanded consistency over style and beauty (both of which Emerson had in abundance). He said much, too, that followed the hopes and biases of the day, and his images often bore a fertile obscurity that allowed many different groups with many different hopes to find many of their own reflections therein, often in exalted form. (It must always be remembered, too, that Emerson wrote speeches not essays; and the literary qualities of performance therefore sometimes win out over clarity of imagery). More than anything, perhaps, he gave then, and continues to give, magnificent voice to the two fundamental attitudes about learning that began in his lifetime and that ruled throughout the century: the yearning for the higher individual as the sum (result and warrant) of higher education; and the Practicalist faith that such education absorb "the living world," such that each single person be allowed his given allotment of greatness, in the name of a better material life and a better Republic.

CHAPTER 4. THE AGE OF JACKSON AND AFTER, PART 2: MYTHS AND MACHINES, IMAGES OF TECHNOLOGY, AND HIGHER LEARNING

1. This is not to say that the new researchers failed in their attempts to gain wider recognition for science. On the contrary, they had several important successes. The founding and setting up of the Smithsonian Institution in Washington, D.C., provides a good example of what the government was willing to do for science at this time. This episode reflects a distinct victory for the new researchers: The Smithsonian was ordained in the name of "research and publication," not solely as a public museum, as many Congressmen wanted (see, e.g., the discussion in Miller, 1970, pp. 3–23, 167–169).

2. In their book on Silliman (which apparently quotes from material since lost), Fulton and Thomson (1947) show the Yale professor's reaction to this incident, as recorded in his diary, to be the following:

> I was not invited to speak but many afterwards expressed regret that I had not been heard. This course appeared singular when they were giving the history of geology in this country and when an individual was present whose life for a half century had been coeval with its origin . . . and whose *Journal of Science* had been the great power wheel which had kept all in motion. . . . It could hardly have been supposed, when some gentlemen spoke who had been instructed by me and most of whom were my personal friends, that my name should have been forgotten. . . . But I do not complain, altho' while I was receiving privately many marks of respect I am unable to explain this public omission. (pp. 256–257)

The blending of felt offense both to personal dignity and to public propriety is entirely at the surface here, in what might otherwise amount to a private confession of complaint, even outrage. Such emotional restraint—put down in a diary, after all—along with the inability to comprehend the larger situation, historically speaking, might be considered very typical of Silliman.

CHAPTER 5. SCIENCE AS "CULTURE": EDUCATION AND MODERNISM IN THE LATE 19TH CENTURY

1. The recapitulation idea, in fact, was a fairly old one by the 19th century, having been proposed in nascent form by Comenius (16th century) and by Vico (early 18th century). Most likely, it is rooted in classical thinking. Certainly, as a biological concept, it had circulated widely during much of the early 19th century in Europe before being given a wholly material evolutionism by Haeckel.

2. During the 1880s, for example, the automobile had become the source of fascination as a "part of the future soon to arrive," and a number of newspapers kept up monthly reports that surveyed and evaluated, in technical language, new prototypes then being tested (see, e.g., Flink, 1975). Such writing, completely absent in today's mass media, would appear to give evidence that keeping up with "progress" entailed a degree of committed self-education in certain technical areas—or, at least a willingness among reporters to become proficient in more than the rhetoric of technology.

3. Harris was possibly the most powerful voice in American secondary education up to the mid-1890s, but thereafter his influence began to erode due to his continued stand for the old curriculum on the basis of the old rationale. His views of education as the denaturalizing of children began to be seen as out of step with the times, reflective of an older, overly authoritarian and poorly informed era. No doubt too, some of Harris's failing popularity was due to his rejection of the natural sciences as important areas for learning. The other grand function of education that he stood for, after all, "to preserve the cultural heritage," he spoke of as the saving of a precious cargo in the midst of a battering storm. That he saw subjects like physics, chemistry, or biology as having no real place on the upper decks of this patrician vessel meant that it had a rather antique, Old World build to it. With the uptake of the sciences into the university and high school curriculum, it was sure to sink, and sink fast (see Harris, 1893).

4. This process of erosion, of unbuilding the traditional curriculum, provides an excellent revelation of Academist values in reverse: Bit by bit, that is, Eliot peeled off each layer of epistemological prestige, until the very core or essence of classicism was left. First to fall to elective status was the *quadrivium* (in effect), subjects such as geography, astronomy, and music. Next came logic and philosophy, then Greek and Latin, and finally, by 1894 (over 20 years after the start), only rhetoric and foreign languages remained. Three years later, the latter became an elective.

5. Shortly after his retirement in 1909 Eliot conceived the so-called Five-Foot Shelf, known today as the "Harvard Classics"—whose 54 volumes he presumed to represent the "greatest thoughts of mankind." He mainly included literary and philosophical authors, such as Homer, Plato, Dante, Shakespeare, and so on. Yet he did not leave out Darwin, Faraday, Pasteur, Jenner, and even Benjamin Franklin. Science too could congratulate itself for having entered the realm of the "great" literati. According to Cremin (1988, p. 385), Collier Press decided to publish the entire set and by some reports sold over 300,000 sets. This figure is probably exaggerated; yet, even if only half the number were purchased, it would indicate the hold that the Academist ideal—the "cultured in-

dividual" as status symbol—continued to have for certain segments of American society (most likely, those with some means).

6. The new importance of science to the government was made clear again a year later, when the National Academy of Sciences was chartered as a separate, "floating" organization to give consultation on technical and military matters during the Civil War. One should note that this idea of creating a "college of tenured experts" to help dictate state policy was one that would recur, time and again, in the early 20th century.

7. By "agriculture," the act did not mean a pastoral, landscape type of farming, reminiscent of earlier decades, but rather a more "scientific" type, involving the latest methods of fertilizing, use of mechanical equipment, crop cycles, soil chemistry, veterinary medicine, and so on.

8. In terms of research and technical learning both, German universities were the envy of the scientific world in Europe and the United States during the entire 19th century. The reasons for their success in this area—most of all, for their concentration on research—were a direct outcome of political intentions.

These schools had undergone a period of radical modernization following the defeat of Prussia by Napolean's army in 1806. In searching out the reasons for this defeat, German political leaders discovered that the French had made excellent use of many technical innovations (better gunpowder, cannon design, roadbuilding techniques, and so on), as well as "scientific strategy" (complex, rationalized battle plans). The decision was made to nurture scientific knowledge as a source of strength and power, and to do this by creating a university system that would act as a generating center of such knowledge. A model was established in the University of Berlin, set up in 1810 by Wilhelm von Humboldt, a well-known philologist (elder brother of Alexander), who the year before had been appointed privy councillor of state in the Prussian government.

Humboldt's idea was to form a synthesis between the older, classics-based university and the active scientific societies that had sprung up during the previous century and that taught science to their members via lectures, demonstrations, discussions, and actual hands-on research. The basic concept was to promote *Wissenschaft*—knowledge in all its forms—as the guiding vision of the curriculum. But a distinct emphasis came to bear on the pursuit of scientific subjects. Selective government funding strongly encouraged this; a series of new institutes and research laboratories, with publishing privileges, scholarship capability, increased salaries, and the most up-to-date and comprehensive equipment, were the result (indeed, even before 1900, the term *Wissenschaft* had itself become synonymous with scientific knowledge). Smaller schools and academies, meanwhile, traditionally operated by local principalities, were closed, and existing institutes were consolidated and raised to the level of universities. A model teaching/research laboratory emerged at the University of Giessen in 1824, under the direction of the already renowned chemist Justus von Liebig.

Within less than decade this laboratory became the acknowledged ideal in the Western world for technical learning at the upper level. Students studied theory and experimental procedure in the classroom, and then applied these directly to their own original research, performed in a fully equipped lab. They were encouraged to publish their work, and, for graduation, were required to prepare a meticulous and well-argued thesis. Standards for written expression were very high, based upon classical Latin models of clarity and rhetorical simplicity. No "crisis" existed between the sciences and humanities; indeed, the latter were seen as necessary to the former.

The basic structural model for performing research became a standard for scientific education later on, especially in the so-called research universities of the United States. Between 1840 and 1880, when such universities began to come into their own, a large number of American scientists went to Germany for some portion of their training.

Nonetheless, German schooling at the preuniversity level remained severely, even rigidly, centered on a medieval-type curriculum. "As in Great Britain, therefore, until 1900 . . . most German scientists of the top rank had had severely classical educations and first experienced science at university" (Brock, 1990, p. 949). Their passing of the difficult *Arbitur* exam automatically conferred elite, "statesman" status upon them. Moreover, the psychology of working in government-funded labs, for a government salary, worked in many to create a sense of loyalty and, at the same time, required diplomacy toward the nation and its people. Many famous German scientists of the latter 19th century were actively engaged in popularization activities, which they did not at all scorn but instead welcomed.

9. The celebration of technology and technological solutions to human questions, however, was an international phenomenon of the highest order, as made plain by Stephen Kern in his book *The Culture of Time and Space: 1880–1918* (1983). He provides a fascinating anecdote that reveals, in emblematic fashion, the general connection between technology, organicism, and progress, in this case centered on the "theory" that electricity should be able to increase the growth of living things:

> Enthusiasm for this theory peaked with the work of the Nobel Prize winning chemist Svante Arrhenius, who tested the effect of electrical stimulation on the growth of children. He placed one group in schoolrooms with wires carrying high-frequency alternating current. After six months the "electrically charged children" had grown twenty millimeters more than those in the control group. The "magnetised teachers" reported that "their faculties were quickened." (p. 114)

CHAPTER 6. SCIENCE AND THE PROGRESSIVES: STANDARDS AND STANDARD BEARERS IN THE AGE OF REFORMISM

1. Positivism, the doctrine that ascribes all "positive," that is, factual and true, knowledge to the methods of science, was founded as a theory by August Comte in the 1830s. Before the arrival of Spencer in the 1860s and 1870s, the theory had no strong spokesperson in America. In the United States, it wasn't until after the acceptance of science into the academy proper—that is, when the discourse of "disinterested inquiry" and "scientific method" became common currency—that positivism as a general intellectual outlook took hold at all levels.

2. Parker's own ideas were drawn directly from earlier European reformers, especially German romantics such as Johann Heinrich Pestalozzi (1746–1827) and Friedrich Froebel (1782–1852). Both these writers saw the education of children, especially young children, as a vast resource for the betterment of humanity. They were dedicated to the idea of improving conditions for the poor. Despite his religious mysticism, Froebel in particular was important to Parker and other American reformers for his ideas regarding play as a necessary educative process in childhood and his emphasis on the idea that learning should be a self-motivated activity.

3. In his program for the Laboratory School Dewey layed out three basic areas of "experience" for learning: work with textiles (sewing, weaving, and so on); cooking activities (growing, planting, harvesting, and the like); and "shop" materials, such as wood and tools. Each of these areas would be introduced to the child on the basis of psychological age groupings, of which there were also three: an early "play period" (4–8 years); a middle period of "spontaneous attention" (8–12 years); and the period of "reflective attention" (12 and up). The first period, in particular, was dedicated to "occupations of the home"; only in the last year was the child introduced to some of the inventions and work of "prehistoric" life, mainly the farm (it is also at this time, the eighth year, when reading and writing are begun). In the second period, a more active discovering of certain inventions and processes, typical of society at a later stage of development, is encouraged. In the final "reflective" period a student is allowed to conceive problems for himself or herself (this program was never worked out, as no students of this age ever attended the Lab School).

An illustration to show how the middle period was to work, at about 10 or 11 years of age, was given by Dewey as follows: The children were given raw harvested materials, such as wool and cotton, and then urged to study them with the goal in mind of making cloth. Wool fibers would appear easier to use, longer and simpler to separate, while cotton would show itself to be more difficult to handle (yet more plentiful, since an agricultural product). Following this, it would be time to "reinvent" the necessary tools for carding the wool (boards with pins or nails sticking out of them), for spinning (a stone or piece of wood with a hole in the middle of it to stretch the fibers out as they are twisted); and so on, with the final endpoint being a crude approximation of an actual loom. The whole process would require a year or more and, along the way, a good deal of history and science would be mixed in. In particular, science would be learned "from the study of the fibres. . . . the conditions under which the raw materials are grown . . . the physics involved in the machinery of production" (Dewey, 1899, p. 45). Similar things could be done with other materials and occupations.

4. Montessori's ideas and methods, it might be noted, introduced into the United States from 1909 onward, were met with harsh and dismissive criticism by a noted proponent of Dewey's ideas, William Kilpatrick, then a professor of education at Teacher's College, Columbia University. Kilpatrick's criticism, offered in his influential book *The Montessori System Examined* (1914), believed a warped appropriation of Dewey by Progressivist thinkers who tended to stress "social cooperation" over individual activity ("individualist" having become, to many, a loaded word connected to laissez-faire capitalism and the "anarchy of greed and self-interest" of the previous generation). Kilpatrick saw "that in the Montessori school the individual child has unusually free rein" (p. 16), and to his mind this was counterproductive to the idea of school as "an embryonic community." Instead, children should be put "into such a socially conditioned environment that they will of themselves spontaneously unite into larger or smaller groups to work out their life impulses" (p. 20). Such comments become important in the larger scheme of how Progressivism as a whole splintered into a host of doctrinaire sects, how its thrust for reform became weakened by factionalism. Kilpatrick, for one, indicates how in the name of "cooperative improvement," many of these factions crashed upon the rock of bad faith in Dewey's "democratic individual," giving authoritarian privilege to forms of social order and "service" over those of private choice, openness, and flexibility.

5. A history of some efforts along these lines, which began during the early 19th century, is given by Simon (1974). For a specific example during this time period (1900–1920), see Teitelbaum, (1991). Though no doubt important from the standpoint of a "critical pedagogy," these attempts at alternative education commonly shared the same view of "science" as that held by the Progressives. As Teitelbaum indicates, the intent was that "a 'scientific attitude' should prevail, in which pupils 'question everything and answer with care' " (p. 154).

6. An interesting example is history itself. The founding of a scientific history involved everything just mentioned. Its ready-made "theory" had been provided by Spencer, and its first home was in the new research university par excellence, Johns Hopkins. Searching out the facts of historical development involved an attempt to identify "the logic of

events." Before long, different universities went ahead and developed their own competitive schools of thought, which could be either wholly Baconian or equally theoretical in their search for "laws" of historical change.

At Johns Hopkins the head of the movement, Herbert Baxter Adams, often gave glory to analogies that ignored the boundaries of metaphor (yet that revealed certain popularizing tendencies). He spoke of his own seminars as being "laboratories where books are treated like mineralogical specimens, passed about from hand to hand, examined, and tested." Scorning all earlier historians, whom he called "amateurs," Adams, like a great many of his contemporaries, argued that history must be written in a dispassionate, uninvolved language centered on "facts" and "principles." This belief, however, did not prevent him from making forays into the wilds of figurative display, as in his claim that America "will yet be viewed and reviewed as an organism of historic growth, developing from minute germs, from the very protoplasm of state life" (in Hofstadter, 1968, p. 39). Animal, vegetable, mineral, or machine, the body of history and of the nation were united by their obedience to science, whether as labor or as nature.

Scientific history enjoyed a brief half-life. By the 1920s it was largely gone as a dominant motif in scholarship, though in style and in the structure of publication it had changed the field permanently. In the late 1910s and 1920s other more famous and influential historians, such as Frederick Jackson Turner and Charles Beard, would follow who were less scientifically minded in their own work yet more concerned with applying the formal understanding of "true" scientists—sociologists, psychologists, political scientists, and economists—to the need for social reform. No longer scientists themselves, historians were still "experts" in their ability to reveal the requirement for expertise.

Chapter 7. What Bearing It May Have: Legacies of Progressivism in the Early 20th Century

1. This is not to say that positivistic faith in science went grandly unopposed between the two world wars. The fundamentalist challenge, mounted with fervor behind William Jennings Bryan during the early 1920s and culminating in the famous Scopes trial (1925), represented a large-scale rejection of "urban" scientific ideas by a sizeable portion of the South in particular. Yet this rejection was aimed almost exclusively at the teachings of geology and evolutionary science, at defending the truth of the Bible and the image of the human being as sacredly separate from all other life. On fields like physics, chemistry, astronomy, or nonevolutionary biology, fundamentalists were largely or completely silent. More than

this, they represented a type of rural rebellion (reminiscent of Bryan's own agrarian movement 30 years before), or rebellion of rural values, in the midst of an ever-more metropolitan culture. If they had any lasting effect on the image of science in America, it was probably to add force to the positivist cause among those who already held power at the national level. It is also interesting to note, in another connection, that the teaching for which Scopes was brought to trial came directly out of a classic product of Progressivist textbook reform—George W. Hunter's *A Civic Biology*— which, having been published in 1914, was based on "biology for better citizenship" ideas, and still more, was itself grossly simplistic and horribly out of date by the mid-1920s, when it became the subject of so much attention and controversy.

2. See Walter Lippman, 1922. Lippmann's ideas came directly from his perception that mainstream "news" had largely become a matter of propaganda, a "manufacturing of consent" imposed not by any central power or agency but by a range of political, economic, and other interest groups bent on manipulating public information for local advantage.

3. Ellen Schrecker (1986), among other historians, has indicated how "racial science" influenced certain aspects of higher education. She notes, for example, that beginning in the 1920s, "when large numbers of smart first- and second-generation East European Jews seemingly threatened to inundate the nation's top colleges, most of those schools followed the lead of Harvard and imposed quotas on the number of Jews they would admit" (p. 30). Furthermore, though Jews were allowed to earn graduate degrees at these schools, they were very rarely hired as professors. Tenured Jews in elite colleges were a veritable unknown, especially in humanities departments; Lionel Trilling became the literary Jackie Robinson when hired by Columbia (his own alma mater) in the 1940s, after having already established a formidible reputation elsewhere. Ironically, "racial science" as a general sensibility may well have urged more Jews to enter the natural sciences (including medicine) as a profession, since it was here that anti-Semitism was somewhat reduced and where the better posts were not in the elite schools but in the research universities.

4. In the beginning technocracy was a decidedly Marxist enterprise, dedicated to the idea of creating a "soviet of engineers" who would aid the working class in their overthrow of the existing system and in instituting an economy run by enlightened "technocrats" (scientists and engineers). Before long, however, the revolutionary side of the equation was dropped (along with Veblen). As the movement gained popularity, it inevitably veered away from overt references to socialism or communism. Its failure, as Segal (1985) partly documents, was due less to the lack of success in this effort than the abrupt and intense swing toward anticollectivist sympathies after 1933 (following, e.g., Hitler's assumption of power, the much-overrated success of the American Communist party in the 1932

elections, and so forth), and, perhaps more importantly, bitter infighting between two groups, one of which was led by "well-to-do cosmopolitans" and opted for wholesale cultural reform, which the other was a gathering of engineers and technicians interested only in economic change.

5. This had come about because of an "experiment" that ended up having enormous historical consequence. In brief, it involved another program combining both engineers and social scientists—one first promoted philosophically by the well-known engineer and educator William Wickenden, who claimed a need for "broader conceptions of the engineer's work in adapting energy and matter to social and economic ends" (in Noble, 1977, p. 318). In the year 1924, this led to a study of lowered productivity at Western Electric's Hawthorne plant, where it became apparent, after 6 full years of analysis, that Taylorite techniques were inadequate to yield an answer to the problem. Instead, the solution came from a group of sociologically minded academics from the Harvard Business School, led by Elton Mayo. This group focused not on management activity but on the workers themselves, whom, they discovered, were the ones who really controlled productivity through a complex social system within the plant, in which a particular worker's overall efficiency rose or fell according to how much personal attention he received from management. This "discovery" helped lead to industrial psychology and "human relations"—but it did something else as well. It showed that "science" was still the supreme educator as far as understanding any kind of "system" was concerned, but that such systems often came down to a more personal level, involving individual feelings and moral choices. This was a mass individual, to be sure, a kind of quantum. But it was nonetheless a psychology far different from that of pure "efficiency" and machine-like "functionality."

6. In blunt terms, the "great ideas" of the Western tradition had played a role in the rise of Hitler and Stalin, and they had done so through the very notion of their perennial and universal validity, by demanding forms of reverence and of obedience that were the very antithesis to the informed and skeptical, revision-minded outlook that Jefferson had said would always be the fount of any effective democracy. After all, Dewey once asked, how many members of Germany's fascist elite were themselves deeply schooled in the great classical tradition? How many of their ideas of leadership, national honor, and destiny might have been acquired or enforced thereby?

7. Before the late 1890s all teaching below the college level was based largely, if not exclusively, on students' repeating in class (aloud or on paper) "lessons" supplied by the teacher. In the sciences, this further involved the taking of notes by the entire class (seated at their desks) while an experiment was demonstrated by the teacher, occasionally assisted by one of the students; on the basis of such notes, a report was

prepared, turned in, and graded. Experiments were listed in an assigned textbook, of which each student was supposed to have a copy. This type of approach was also used in many colleges and was based on the notion that the teacher/professor should demonstrate, if nothing else, proper laboratory procedure, and the need for exact observation and description. It was also followed for another, more practical reason. Scientific equipment was expensive, and most schools did not have enough money to outfit labs where students themselves could do work individually or in small groups.

CHAPTER 8. THE POSTWAR ERA: THE RETURN OF ACADEMISM AND THE SPUTNIK "REVOLUTION"

1. All quotes from *The University of Utopia* are taken from the 1964 second edition.

2. The "two cultures" debate, in which the natural sciences and the humanities were pitted against one another, might be defined as a conflict between two different sects of Academism, two different models for elite leadership. It derives its title from C. P. Snow's famous Rede Lecture at Cambridge, published as an essay under that title in the *New Statesman*, October 6, 1956 (in Snow, 1963). Snow claimed precedence for science as an "intellectual force in the modern world" and chastised the humanities for their abject ignorance and rejection of it. In fact, he was speaking an ancient refrain, of which he himself was largely ignorant.

As a historical phenomenon, that is, the debate stands at the head of a very old and continuous controversy, one that has nearly always involved "science" in one form or another, and which has been focused more generally on the relative value of "ancient" wisdom and "modern" knowledge. The argument began in classical antiquity, became a source of much furor during the 13th century, with the introduction of Aristotlean thought (marking the fall of the old "liberal arts") and the rise of the medieval university, and reappeared during the late 17th and early 18th centuries as the famed "quarrel of the ancients and moderns," waged more directly either for or against the idea of "progress" (mainly associated with scientific knowledge). In the 19th century, it resurfaced as a debate between science and art among the romantics, and later, in the Victorian period, between science and literature (as in the rejections of science sponsored by those such as John Ruskin and Mathew Arnold). On the plains of institutional power, it was reflected in the conflict over whether or not to allow scientific studies in the colleges, both in England and America (in France and Germany, they had already become integrated).

Never having gone away, the debate came back in force during the era of positivist hegemony, the 1920s to the 1950s, when it was becoming especially clear that the humanities had lost enormous prestige in the academy. At this point, therefore, controversy was an expression of a growing shift/struggle in power relations matching that in the larger society. Much of the resulting conflict was focused on each side trying to claim essentialist territory for its own prestige, for example, literature as sacred habitat for "the life of the imagination," science as the abode of "action" and "the future." Though the debate was never resolved—indeed, literature and scientific departments remain the most physically separated on every college campus—it lost steam by the late 1960s, when the authority-based claims of both sides came under heavy criticism by those for whom the controversy was presumably waged (the students). For several well-known works that sum up the relevant points of view, see Richards (1935) and Snow (1963). For responses to both sides, see Leavis (1963).

3. These statements should be qualified somewhat. Ralph Tyler has had a long, varied, and fertile career in education and his contributions have vastly exceeded the rationale discussed. The long-term power granted his early ideas had everything to do with that portion of the educational establishment that has remained interested in technological solutions to basic pedagogic problems.

Bloom, on the other hand, is among the most influential educational psychologists of the last 40 years, and went on to help give advice to the Head Start program, initiated in the summer of 1965. One of his abiding interests has been the influence of the family on intellectual development. In the 1980s, for example, he published work showing the importance of family influence and support in the careers of successful scientists and musicians, implying that "talent" or "genius" were not, as many believed, the determining factor in success.

CHAPTER 9. NEW VOICES AND OLD LIMITS: REFORMISM IN THE LATE 1960s

1. Though student protest and political movements associated with left-wing ideas had existed before, most notably during the 1930s, nothing of the scale of the 1960s had ever been witnessed in the United States. Indeed, up to this time, one of the most distinguishing features of the American university relative to Western culture as a whole was that it had never become the generating center of political revolt. Such revolt throughout the 19th and 20th centuries, by contrast, was often begun in the universities of Paris, Vienna, Berlin, Prague, Rome, Moscow, Madrid, and so forth.

2. Of all the programs begun during this period, Head Start is the one perceived to work best by observers of all political persuasions. It is therefore the one to have survived the longest.

3. As it happens, there are undoubtedly a number of reasons for this decline, whose meaning is therefore entirely unclear. One likely cause,

however, has to do with increased competition for getting into college, a factor which, as Goodlad (1984) states, has encouraged students to more often apply in their junior year of high school, thus submiting test scores from this year and not the senior year.

4. All this, perhaps, should be added to the general context of claims about the environmental movement as a whole, particularly those that see it as representing, both in the United States and Europe, the most successful form of oppositional politics to have emerged out of the late 1960s.

5. One might note, for example, that Kuhn (1962) was ironically published as a volume in the International Encyclopedia of Unified Science series, under the editorial direction of Otto Neurath and Rudolf Carnap, two of the most adamant proponents of the old positivism.

6. I don't mean to imply that Kuhn (1962) was universally read within the scientific community. The proportion of those who gained any significant familiarity with his book were probably few. Yet such terms as "paradigm," "normal science," and, of course, "revolution," became shared catchwords of the day by the late 1970s. Indeed, among certain fields, such as geology, which all agreed had undergone a "paradigm shift" with the introduction of plate tectonics, Kuhn was so often invoked (and so little read) as to become a tired cliché.

7. In point of fact, sociology of science had arrived as early as the 1940s, with Robert K. Merton as its pioneer. Merton, who belonged to the functionalist school of sociology (led by Talcott Parsons), sought to examine how science "operated" in terms of institutional norms. These norms were viewed as fixed rules of the game, the guiding principles that directed scientists to act in ways that would maintain the character and quality of technical knowledge. These norms were: communality, universalism, organized skepticism, and disinterestedness. Scientists were urged by these codes to judge new evidence and ideas on their intellectual merits alone, untouched by external factors. Studies in the history of science were said to reveal how conformity to these codes had kept scientific knowledge objective (see, e.g., Merton, 1939). Mertonian sociology of science was therefore a strictly positivist affair, an attempt to show how science worked as a scientific system, separated from the rest of society by its internal "functions." Such views dominated the field throughout the 1950s and 1960s. In a sense, the explosion of science studies thereafter might be seen as a release of intellectual energies out from under the gross professional weight that Mertonian sociology had come to acquire.

8. A few recent works in this general category of curriculum critique are Apple (1979, 1986); Apple and Weis, eds. (1983); Whitty (1985); Apple and Christian-Smith, eds. (1991).

Chapter 10. Troubled Symbolisms: Science and the Curriculum at the Century's End

1. For a good, brief example of this kind of claims making (put before a professional audience), see Grogan (1990).

2. The published goals of the new NSF effort are as follows: "(i) raising the level of excellence and standards of student performance . . . ; (ii) developing new curricula and course materials . . . ; (iii) raising the qualifications of precollege teaching faculty; (iv) stimulating interest in science and mathematics faculty concerned with precollege education; (vi) increasing the numbers of graduate students" (Bloch, 1990, pp. 839–840). Nowhere in the entire grouping of goals and programs offered by this new campaign, it might be noted, is the possibility of low-interest loans to potentially interested students mentioned.

3. For example, in the section concerning geologic science (my own particular field), students are encouraged to learn how to identify common rocks and minerals; to construct a geologic time scale showing "how little of geologic time humans . . . have occupied" (National Science Teachers Association, 1992, p. 77); to classify sedimentary rocks; to study fossils; to assemble the continents as a "jigsaw puzzle"; and so on. These are time-worn practices that have graced secondary classrooms for decades.

4. A brief list of such studies might include the following: Champagne and Hornig (1987), Mullis and Jenkins (1988), Weiss (1987), and Holdzkom and Lutz (1985).

References

Abir-Am, P., & Outram, D. (Eds.). (1987). *Uneasy careers and intimate lives: Women in science, 1789–1979*. New Brunswick, NJ: Rutgers University Press.

Adams, H. (1885). *History of the United States of America during the administrations of Thomas Jefferson*. New York: Library of America, 1986.

Aldridge, B. G. (1992). Essential changes in secondary science: Scope, sequence, and coordination. In M. K. Pearsall (Ed.), *Scope, sequence, and coordination of secondary school science, Volume II: Relevant research*. Washington, DC: National Science Teachers Association.

Allen, F. L. (1931). *Only yesterday: An informal history of the twenties*. New York: Harper & Row, 1964.

Allen, G. S. (1922, Winter). Master mechanics and evil wizards: Science and the American imagination from Frankenstein to Sputnik. *Massachusetts Review, 33*(4), 505–558.

Apple, M. W. (1979). *Ideology and curriculum*. London: Routledge.

Apple, M. W. (1986). *Teachers and texts: A political economy of class and gender relations in education*. New York: Routledge.

Apple, M. W. & Christian-Smith, L. K. (1991). *The politics of the textbook*. New York: Routledge.

Apple, M. W., & Weiss, L. (Eds.). (1983). *Ideology and practice in schooling*. Philadelphia: Temple University Press.

Association of American Colleges (1985). *Integrity in the college curriculum: A report to the academic community*. Washington, DC: Association of American Colleges.

Bailey, S. K. (1981, Summer). Political coalitions for public education. *Daedalus, 10*, 27–43.

Bailyn, B. (1960). *Education in the forming of American society*. New York: Norton.

Bailyn, B. (1967). *The ideological origins of the American revolution*. Cambridge, MA: Harvard University Press.

295

Ballantine, J. (1983). *The sociology of education.* Englewood Cliffs, NJ: Princeton University Press.

Barnes, B. (1990). Sociological theories of scientific knowledge. In R. C. Olby, G. N. Cantor, J. R. R. Christie, & M. J. S. Hodge (Eds.), *Companion to the history of modern science* (pp. 60–73). London: Routledge.

Bayertz, K. (1985). Spreading the spirit of science: Social determinants of the popularization of science in 19th-century Germany. In T. Shinn & R. Whitley (Eds.), *Expository science: Forms and functions of popularization* (pp. 209–227). Boston: D. Reidel.

Beaver, D. De B. (1971). Altruism, patriotism, and science: Scientific journalism in the early republic. *American Studies, 12,* 10.

Bell, W. J. Jr. (1955). *Early American science: Needs and opportunities for study.* Chapel Hill: University of North Carolina Press.

Bennett, W. J. (1984). *To reclaim a legacy: A report on the humanities in higher education.* Washington, DC: National Endowment for the Humanities.

Berger, P. (1967). *The sacred canopy.* Garden City: Doubleday.

Bigelow, J. (1829). *Elements of technology.* Boston: Boston Press.

Bloch, E. (1990). Education and human resources at the National Science Foundation. *Science, 247,* 839–840.

Bloom, A. (1987). *The closing of the American mind.* New York: Simon & Schuster.

Bloom, B. S. (1956). *Taxonomy of educational objectives.* New York: John Wiley.

Bloom, B. S. (1964). *Stability and change in human characteristics.* New York: John Wiley.

Bloom, B. S. (1981). *All our children learning.* New York: McGraw-Hill.

Bode, C. (1956). *The American lyceum: Town meeting of the mind.* New York: Oxford.

Boenig, R. W. (1969). *Research in science education: 1938 through 1947.* New York: Teachers College Press, Columbia University.

Boller, P. F. Jr. (1969). *American thought in transition: The impact of evolutionary naturalism, 1865–1900.* Chicago: Rand McNally.

Boorstin, D. J. (1966). *The Americans: The national experience.* New York: Random House.

Bowen, J. (1975). *A history of western education, Vol. II: Sixth to sixteenth century.* New York: St. Martin's Press.

Boyd, W. (1966). *The history of western education* (8th ed.). New York: Barnes & Noble.

Boyer, E. L. (1987). *College: The undergraduate experience in America.* New York: Harper & Row.

Brock, W. H. (1990). Science education. In R. C. Olby, G. N. Cantor, J. R. R. Christie, & M. J. S. Hodge (Eds.), *Companion to the history of modern science* (pp. 946–960). London: Routledge.

Brown, C. M. (1989). *Benjamin Silliman: A life in the young republic.* Princeton: Princeton University Press.

Brown, J. (1975). *A history of western education, Vol. II: Sixth to sixteenth century.* New York: St. Martin's Press.

Bruce, R. V. (1987). *The launching of modern American science 1846–1876.* New York: Knopf.

Bruner, J. (1960). *The process of education.* Cambridge, MA: Harvard University Press.

Calvert, K. (1992). *Children in the house: The material culture of early childhood, 1600–1900.* Boston: Northeastern University Press.

Carr, H. Y., & Weidner, R. T. (1971). *Physics from the ground up.* New York: McGraw-Hill.

Carter, P. A. (1977). *The creation of tomorrow: Fifty years of magazine science fiction.* New York: Columbia University Press.

Cavenagh, F. A. (Ed.). (1932). *Herbert Spencer on education.* Cambridge, England: Cambridge University Press.

Champagne, A., & Hornig, L. (Eds.). (1987). *Students and science learning: Papers from the 1987 National Forum for School Science.* Washington, DC: American Association for the Advancement of Science.

Christie, J. R. R. (1990). Feminism and the history of science. In R. C. Olby, G. N. Cantor, J. R. R. Christie, & M. J. S. Hodge (Eds.), *Companion to the history of modern science* (pp. 100–101). London: Routledge.

Cobban, A. B. (1975). *The medieval universities: Their development and organization.* London: Methuen.

Cole, S. (1938). *Liberal education in a democracy.* New York: Harper & Bros.

Conant, J. B. (Ed.). (1945). *General education in a free society.* Cambridge, MA: Harvard University Press.

Conant, J. B. (1952). *Modern science and modern man.* New York: McGraw-Hill.

Conant, J. B. (1953). *Education and liberty: The role of schools in a modern democracy.* Cambridge, MA: Harvard University Press.

Conant, J. B. (1957). Foreword. In T. S. Kuhn, *The Copernican revolution* (pp. xiii–xviii). New York: Vintage Books.

Conant, J. B. (1959). *The American high school today: A first report to interested citizens.* New York: McGraw-Hill.

Conant, J. B. (Ed.). (1959). *Harvard case studies in the history of science.* Cambridge, MA: Harvard University Press.

Conant, J. B. (1961). *Slums and suburbs: A commentary on schools in metropolitan areas.* New York: McGraw-Hill.

Conant, J. B., & Spaulding, F. (1940). *Education for a classless society.* Cambridge, MA: Harvard University Press.

Counts, G. S. (1932). *Dare the school build a new social order?* New York: John Day (No. 11 John Day Pamphlets).

Cremin, L. A. (Ed.). (1957). *The republic and the school: Horace Mann on the education of free men.* New York: Teachers College Press, Columbia University.

Cremin, L. A. (1970). *American education: The colonial experience—1607–1783.* New York: Harper & Row.

Cremin, L. A. (1977). *Traditions of American education.* New York: Basic Books.

Cremin, L. A. (1980). *American education: The national experience—1783–1876.* New York: Harper & Row.

Cremin, L. A. (1988). *American education: The metropolitan experience—1876–1980.* New York: Harper & Row.

Cremin, L. A. (1989). *Popular education and its discontents.* New York: Harper & Row.

Cuban, L. (1984). *How teachers taught: Constancy and change in American classrooms, 1890–1980.* New York: Longman.

Cubberley, E. P. (1909). *Changing conceptions of education.* Boston: Houghton Mifflin.

Cubberley, E. P. (1916). *Public school administration.* Boston: Houghton Mifflin.

Curtius, E. R. (1953). *European literature and the Latin middle ages* (W. Trask, Trans.). New York: Bollingen.

Daniels, G. H. (1967). The pure science ideal and democratic culture. *Science, 156,* 1699–1705.

Daniels, G. H. (1968). *American science in the age of Jackson.* New York: Columbia University Press.

Day, J., & Kingsley, J. (1829). Original papers in relation to a course of liberal education. *American Journal of Science and Arts, 15*(2), 297–351.

deMause, L. (Ed.). (1974). *The history of childhood.* New York: Harper & Row.

Dewey, J. (1897). My pedagogic creed. *Journal of the National Education Association, 18,* December 1929 (rev. ed.).

Dewey, J. (1899). *The school and society* (rev. ed.). Chicago: University of Chicago Press, 1990 (republication of 1915 rev. ed).

Dewey, J. (1901). The situation as regards the course of study. *Journal of Proceedings and Addresses, 40th Annual Meeting, National Education Association,* 337–338.

Dewey, J. (1902). *The child and the curriculum.* Chicago: University of Chicago Press, 1990.

Dewey, J. (1910). *How we think.* Boston, MA: Heath.

Dewey, J. (1916). *Democracy and education.* New York: Capricorn.

Dewey, J. (1936). Review of Robert Hutchins *The higher learning in America. Social Frontier, 3,* 104.

Dewey, J. (1939). *Freedom and culture*. New York: Capricorn.

Duschl, R. A. (1990). *Restructuring science education*. New York: Teachers College University Press.

Dworkin, M. (Ed.). (1959). *Dewey on education*. New York: Teachers College University Press.

Elias, R. H. (1973). *Entangling alliances with none: An essay on the individual in the American twenties*. New York: Norton.

Emerson, R. W. (1982). Selections from *Journals*. In J. Porte (Ed.), *Emerson in his journals*. Cambridge, MA: Belknap, Harvard University Press.

Emerson, R. W. (1983). In J. Porte (Ed.), *Essays and Lectures*. New York: Library of America.

Fitzpatrick, J. C. (Ed.). (1940). *The writings of George Washington*. Washington, DC: U.S. Government Printing Office.

Fitzpatrick, F. L. (Ed.). (1960). *Policies for science education*. New York: Teachers College University Press.

Flink, J. J. (1975). *The car culture*. Cambridge, MA: MIT Press.

Forgan, S. (1989). The architecture of science and the idea of a university. *Studies in history and philosophy of science, 20*(4), 405–434.

Frank, J. (1935). *Law and the modern mind*. New York: Brentano's.

Franklin, B. (1778). Constitution of Phillips Academy. In T. R. Sizer (Ed.), *The age of the academies* (pp. 77–79). New York: Teachers College, Columbia University, 1964.

Fuller, S. (1989). *Philosophy of science and its discontents*. Boulder, CO: Westview. (2nd ed., 1993. New York: Guilford Press).

Fulton, J. F., & Thomson, E. H. (1947). *Benjamin Silliman: Pathfinder in American science*. New York: Henry Schuman.

Geiger, R. L. (1986). *To advance knowledge: The growth of American research universities, 1900–1940*. New York: Oxford.

Gilman, D. C. (1898). *University problems in the United States*. New York: Century Publications.

Glacken, C. J. (1967). *Traces on the Rhodian shore: Nature and culture in western thought*. Berkeley: University of California Press.

Gomory, R. E., & Shapiro, H. T. (1988, Summer). A dialogue on competitiveness. *Issues in Science and Technology, 4*(4), 36–42.

Goodlad, J. I. (1975). *The dynamics of educational change*. New York: McGraw-Hill.

Goodlad, J. I. (1984). *A place called school: Prospects for the future*. New York: McGraw-Hill.

Goodman, P. (1962). *The community of scholars*. New York: Random House.

Goodman, P. (1964). *Compulsory mis-education*. New York: Horizon.

Goodman, P. (1968). *People or personnel* and *Like a conquered province*. New York: Vintage Books.

Graff, H. J. (1987). *The legacies of literacy.* Bloomington: Indiana University Press.

Graubard, A. (1972). *Free the children: Radical reform and the free school movement.* New York: Pantheon.

Greene, J. C. (1984). *American science in the age of Jefferson.* Ames: Iowa State University Press.

Grogan, W. R. (1990). Engineering's silent crisis. *Science, 247,* 381.

Guralnick, S. M. (1975a). *Science and the ante-bellum American College.* Philadelphia: American Philosophical Society.

Guralnick, S. M. (1975b). Sources of misconception on the roles of science in the nineteenth-century American college. *ISIS, 65,* 352–366.

Gutman, H. G. (1977). *Work, culture, and society.* New York: Knopf.

Hansen, A. O. (1926). *Liberalism and education in the eighteenth century.* New York: Columbia University Press.

Harris, W. T. (1893). *The theory of education.* Syracuse: C. W. Bardeen.

Harris, W. T. (1898). *Psychological foundations of education: An attempt to show the genesis of the higher faculties of the mind.* New York: D. Appleton & Co.

Harvard Committee on the Objectives of a General Education. (1945). *General education in a free society.* Cambridge, MA: Harvard University Press.

Hawke, D. F. (1988). *Nuts and bolts of the past.* New York: Harper & Row.

Hawthorne, N. (1982). In R. H. Pearce (Ed.), *Tales and sketches.* New York: Library of America.

Hindle, B. (1956). *The pursuit of science in revolutionary America, 1735–1789.* Chapel Hill: University of North Carolina Press.

Hofstadter, R. (1963). *Anti-intellectualism in American life.* New York: Knopf.

Hofstadter, R. (1968). *The progressive historians* (rev. ed.). New York: Vintage Books, 1970.

Hofstadter, R., & Metzger, W. P. (1955). *The development of academic freedom in the United States.* New York: Columbia University Press.

Hofstadter, R., & Smith, W. (Eds.). (1961). *American higher education: A documentary history* (Vol.s 1 & 2). Chicago: University of Chicago Press.

Holdzkom, D., & Lutz, P. (Eds.). (1985). *Research within reach: Science education.* Washington, DC: National Science Teachers Association.

Holmes Group. (1986). *Tomorrow's teachers.* East Lansing, MI: Author.

House, E. (1974). *The politics of educational innovation.* Berkeley: McCutchan.

Hughes, T. P. (1989). *American genesis: A century of invention and technological enthusiasm.* New York: Penguin.

Hunter, G. W. (1914). *A civic biology.* New York: American Book Company.

Hutchins, R. M. (1936). *Higher learning in America.* New Haven: Yale University Press.

Hutchins, R. M. (1953). *The university of Utopia.* Chicago: University of Chicago Press (2nd ed., 1964).

Hutchinson, C. E., & Muller, C. B. (1988, Summer). Educating engineers: In praise of diversity. *Issues in science and technology, 4*(4), 71–74.

Illich, I. (1970). *Deschooling society.* New York: Harper & Row.

Irving, W. (1809). A history of New York, from the beginning of the world to the end of the Dutch dynasty. In J. W. Tuttleton (Ed.), *History, tales and sketches.* New York: Library of America, 1983.

James, W. (1890). *The principles of psychology* (Vols. 1 & 2). New York: Henry Holt.

Jarves, J. J. (1864). *The art-idea.* Boston: Hurd & Houghton, 1960 (Cambridge: Belknap, Harvard University Press, Rowland, B. Jr. Ed.).

Jean, F. C., Harrah, E. C., & Herman, F. L. (1934). *An introductory course in science for colleges.* Boston: Athenaeum.

Jefferson, T. (1984). In M. D. Peterson (Ed.), *Writings.* New York: Library of America.

Jencks, C., & Riesman, D. (1968). *The academic revolution.* New York: Doubleday.

Jenkins, L. B., & MacDonald, W. B. (1989, Spring). Science teaching in the spirit of science. *Issues in science and technology, 5*(3), 60–65.

Johnson, S. (1755). *A dictionary of the English language* (Vols. 1 & 2). London: William Strahan.

Kaestle, C. F. (1983). *Pillars of the republic: Common schools and American society, 1780–1860.* New York: Hill & Wang.

Kahle, J. B. (1990). Why girls don't know. *What research says to the science teacher—The process of knowing, 6,* 55–67.

Karier, C. J. (1986). *Scientists of the mind: Intellectual founders of modern psychology.* Urbana, IL: University of Illinois Press.

Kasson, J. F. (1976). *Civilizing the machine: Technology and republican values in America, 1776–1900.* New York: Grossman (Viking).

Katz, M. B. (1971). *Class, bureaucarcy, and schools: The illusion of educational change in America.* New York: Praeger.

Keller, E. F. (1985). *Reflections on gender and science.* New Haven: Yale University Press.

Kern, S. (1983). *The culture of time and space: 1880–1918.* Cambridge, MA: Harvard University Press.

Kerr, C. (1963). *The uses of the university.* Cambridge, MA: Harvard University Press.

Kilpatrick, W. (1914). *The Montessori system examined.* Boston: Houghton Mifflin.

Kliebard, H. M. (1992). *Forging the American curriculum: Essays in curriculum history and theory.* New York: Routledge.

Knight, E. W., & Hall, C. L. (1951). *Readings in American educational history.* New York: Appleton-Century-Crofts.

Kohl, H. (1969). *The open classroom.* New York: Random House.

Kohler, R. E. (1991). *Partners in science: Foundations and natural scientists, 1900–1945.* Chicago: University of Chicago Press.

Kohlsted, S. G. (1976). *The formation of the American scientific community: The American association for the advancement of science, 1848–1860.* Urbana, IL: University of Illinois Press.

Kohlsted, S. G. (1990). Parlors, primers, and public schooling: Education for science in 19th-century America. *ISIS, 81,* 425–445.

Kuhn, T. (1962). *The structure of scientific revolutions.* Chicago: University of Chicago Press, 1970.

Kuritz, H. (1981). The popularization of science in 19th-century America. *History of Education Quarterly, 21,* 259–274.

Kuznick, P. (1987). *Beyond the laboratory: Scientists as political activists in 1930s America.* Chicago: University of Chicago Press.

Labaree, L. W. (Ed.). (1959). *The papers of Benjamin Franklin* (Vol. 3, pp. 422–428). New Haven: Yale University Press.

Larkin, J. (1988). *The reshaping of everyday life: 1790–1840.* New York: Harper & Row.

Laudan, L. (1990). The history of science and the philosophy of science. In R. C. Olby, G. N. Cantor, J. R. R. Christie, & M. J. S. Hodge (Eds.), *Companion to the history of modern science* (pp. 47–59). London: Routledge.

Lawrence, E. S. (1970). *The origins and growth of modern education.* Harmondsworth: Penguin.

Leavis, F. R. (Ed.). (1963). *Two cultures: The significance of C. P. Snow.* New York: Pantheon.

Levine, L. (1988). *Highbrow/lowbrow: The education of T. C. Mitts.* New York: Norton.

Lindberg, D. C. (1992). *The beginnings of western science: The European scientific tradition in philosophical, religious, and institutional context.* Chicago: University of Chicago Press.

Lippmann, W. (1992). *Public opinion.* New York: Harcourt, Brace.

Livingstone, D. W. (1987). Introduction. In D. W. Livingstone & Contributors (Eds.), *Critical pedagogy and cultural power* (pp. 1–15). South Hadley, MA: Bergin & Garvey.

Luke, A. (1988). *Literacy, textbooks, and ideology.* Philadelphia: Falmer Press.

Lurie, E. (1960). *Louis Aggasiz: A life in science.* Chicago: University of Chicago Press.

MacLure, W. (1817). Selections from *Observations on the geology of the United States.* In K. F. Mather & S. L. Mason (Eds.), *A source book in geology, 1400–1900* (pp. 168–174). Cambridge, MA: Harvard University Press, 1970.

Man and learning in modern society. (1959). Papers and addresses delivered at the Inauguration of Charles E. Odegaard as President of the University of Washington. Seattle: University of Washington Press.

Mann, H. (1837–1848). Selections from his reports to the Massachusetts Board of Education. (See Cremin, 1957.)

Mann, M. P. (1865). *Life of Horace Mann.* Boston: Walker, Fuller.

Marx, L. (1964). *The machine in the garden: Technology and the pastoral ideal in America.* New York: Oxford.

Mather, C. (1705). Some special points, relating to the education of my children. In P. Miller & T. H. Johnson (Eds.), *The Puritans: A sourcebook of their writings* (pp. 724–725). New York: Harper Torch Books, 1963.

Mayfield, J. (1982). *The New Nation: 1800–1845.* New York: Hill & Wang.

Mayhew, K., & Edwards, A. C. (1936). *The Dewey school.* New York: D. Appleton-Century.

McDougall, W. A. (1985). . . . *the heavens and the Earth: A political history of the space age.* New York: Basic Books.

Merton, R. K. (1939). *Science, technology, and society in 17th century England.* New York: Free Press.

Miller, H. S. (1970). *Dollars for research: Science and its patrons in 19th-century America.* Seattle: University of Washington Press.

Miller, P. (1964). *Errand into the wilderness.* New York: Harper & Row.

Miller, P., & Johnson, T. H. (Eds.). 1963. *The Puritans: A sourcebook of their writings* (Vol. 2). New York: Harper Torch Books.

Morrison, K. F. (1983). Incentives for studying the liberal arts. In D. L. Wagner (Ed.), *The seven liberal arts in the middle ages.* Bloomington: Indiana University Press.

Moscovici, S. (1985). *The age of the crowd* (J. C. Whitehouse, Trans.). Cambridge, England: Cambridge University Press.

Mullis, I. V. S., & Jenkins, L. B. (1988). *The science report card: Elements of risk and recovery.* Princeton: Educational Testing Service.

Murnane, R. J., & Raizen, S. A. (Eds.). (1988). *Improving indicators of the quality of science and mathematics educations in grades K–12.* Washington, DC: National Research Council.

National Commission on Excellence in Education (1983). *A nation at risk.* Washington, DC: U.S. Government Printing Office.

National Education Association. (1893). *Report of the committee on secondary school studies.* Washington, DC: U.S. Government Printing Office.

National Education Association. (1911). Report of the Committee of Nine on the articulation of high school and college. *NEA Proceedings,* 555–585.

National Education Association. (1918). *Cardinal principles of secondary education: A report of the commission on the reorganization of secondary education.* Washington, DC: U.S. Government Printing Office.

National Endowment for the Humanities. (1987). *American memory: A report on the humanities in the nation's public schools.* Washington, DC: U.S. Government Printing Office.

National Endowment for the Humanities. (1989). *50 hours: A core curriculum for college students.* Washington, DC: U.S. Government Printing Office.

National Science Teachers Association. (1992). *Scope, sequence and coordination of secondary school science, Vol. I: The content core, a guide for curriculum designers.* Washington, DC: Author.

Neill, A. S. (1960). *Summerhill.* New York: Hart.

Nelkin, D. (1977). *Science textbook controversies and the politics of equal time.* Cambridge, MA: MIT Press.

Nelkin, D. (1987). *Selling science: How the press covers science and technology.* New York: W. H. Freeman.

Nietzsche, F. (1887). *Zur genealogie der moral.* Berlin: Goldman, 1983, P. Putz (Ed.).

Noble, D. F. (1977). *American by design: Science, technology, and the rise of corporate capitalism.* New York: Knopf.

North American Review. (1855). European and American Universities, 80, 150–155.

Novak, B. (1980). *Nature and culture: American landscape and painting 1825–1875.* London: Thames & Hudson.

Oakes, J. (1985). *Keeping track: How schools structure inequality.* New Haven: Yale University Press.

Oakley, F. (1992). *Community of learning: The American college and the liberal arts tradition.* New York: Oxford.

Office of Technology Assessment, U.S. Congress. (1988). *Elementary and secondary education for science and engineering: A technical memorandum.* Washington, DC: U.S. Government Printing Office.

Parker, F. W. (1895). Contribution to the discussion of Dr. C. C. Van Liew's essay on "Culture Epochs," National Society for the Study of Education first supplement. *National Herbart Society Yearbook,* 155–157.

Pauly, P. J. (1991). The development of high school biology: New York City, 1900–1925. *ISIS, 82,* 662–668.

Pearsall, M. K. (Ed.). (1992). *Scope, sequence, and coordination of secondary school science, Vol. II: Relevant research.* Washington, DC: National Science Teachers Association.

Peterson, M. D. (Ed.). (1984). *Writings.* New York: Library of America.

Polyani, M. (1946). *Science, faith and society.* London: Oxford.

Porter, C. M. (1986). *The eagle's nest: Natural history and American ideas, 1812–1842.* Tuscaloosa: University of Alabama Press.

Porter, R. (1990). The history of science and the history of society. In R. C. Olby, G. N. Cantor, J. R. R. Christie, & M. J. S. Hodge (Eds.). *Companion to the history of modern science* (pp. 32–46). London: Routledge.

Pursell, C. W. Jr. (1981). *Technology in America: A history of individuals and ideas.* Cambridge, MA: MIT Press.

Ravetz, J. R. (1990). Orthodoxies, critiques, and alternatives. In R. C. Olby, G. N. Cantor, J. R. R. Christie, & M. J. S. Hodge (Eds.), *Companion to the history of modern science* (pp. 898–909). London: Routledge.

Reingold, N. (1991). *Science, American style.* New Brunswick, NJ: Rutgers University Press.

Richards, I. A. (1935). *Science and poetry.* New York: Norton.

Rice, J. M. (1893). *The public school system of the United States.* New York: Century.

Rippa, S. A. (1992). *Education in a free society* (7th ed.). New York: Longman.

Robinson, G. (1979). Edward Hitchcock. In L. G. Wilson (Ed.), *Benjamin Silliman and his circle* (pp. 49–84). New York: Science History Publications.

Rogers, D. T. (1985). Socializing middle-class children: Institutions, fables, and work values in 19th-century America. In N. R. Hiner & J. M. Hawes (Eds.), *Growing up in America: Children in historical perspective* (pp. 119–135). Chicago: University of Illinois Press.

Rossiter, M. W. (1982). *Women scientists in America: Struggles and strategies to 1940.* Baltimore: Johns Hopkins.

Rothman, D. J. (1971). *The discovery of the asylum: Social order and disorder in the New Republic.* Boston: Little, Brown.

Rouse, J. (1991). Philosophy of science and the persistent narratives of modernity. *Studies in the history and philosophy of science, 22*(1), 141–162.

Rudoph, F. (1977). *Curriculum: A history of the undergraduate course of study since 1636.* San Francisco: Jossey-Bass.

Runes, D. D. (Ed.). (1947). *The selected writings of Benjamin Rush.* New York: Philosophical Library.

Rutherford, F. J., & Ahlgren, A. (1990). *Science for all Americans.* New York: Oxford.

Schlesinger, A. M. Jr. (1992). *The disuniting of America: Reflections on a multicultural society.* New York: Norton.

Schrecker, E. W. (1986). *No ivory tower: McCarthyism and the universities.* New York: Oxford.

Searle, J. (1990, December 6). The storm over the university. *New York Review of Books, 37*(19), 34–42.

Segal, H. P. (1985). *Technological utopianism in American culture.* Chicago: University of Chicago Press.

Shapin, S. (1990). Sciences and the public. In R. C. Olby, G. N. Cantor, J. R. R. Christie, & M. J. S. Hodge (Eds.), *Companion to the history of modern science* (pp. 990–1008). London: Routledge.

Sienko, M. J., & Plane, R. A. (1966). *Chemistry: Principles and properties.* New York: McGraw-Hill.

Sillman, B. (1819). "Plan of the work" and "Introductory remarks." *American Journal of Science and Arts, 1*(1), v–vi, 1–8.

Simon, B. (1974). *Studies in the history of education* (3 Vols.). London: Lawrence & Wishart.

Simpson, S. (1965). The working man's manual. Selections given in R. Vassar (Ed.), *Social history of American education: Vol. 1. Colonial times to 1860* (pp. 182–194). Chicago: Rand McNally. (Original work published 1831)

Smith, M. R. (1985). *Military enterprise and technology change.* Cambridge, MA: MIT Press.

Smith, S. S. (1787). *An essay on the causes of the variety of complexion and figure in the human species.* Philadelphia: Robert Aiken.

Snow, C. P. (1963). *The two cultures: And a second look.* New York: Mentor.

Spencer, H. (1850). *Social statics; or the conditions essential to human happiness, specified, and the first of them developed.* New York: D. Appleton, 1882.

Spencer, H. (1860). *Education: Intellectual, moral and physical.* New York: D. Appleton, 1871.

Spring, J. (1976). *The sorting machine: National educational policy since 1945.* New York: Longman.

Spring, J. (1989). *American education: An introduction to social and political aspects* (4th ed.). New York: Longman.

Stroud, P. T. (1992). *Thomas Say: New world naturalist.* Philadelphia: University of Pennsylvania Press.

Sutherland, D. E. (1989). *The expansion of everyday life: 1860–1876.* New York: Harper & Row.

Teitelbaum, K. (1991). Critical lessons from our past: Curricula of Socialist Sunday schools in the United States. In M. Apple & L. K. Christian-Smith (Eds.), *The politics of the textbook* (pp. 135–165). New York: Routledge.

Thorndike, E. L. (1912). *Education: A first book.* New York: Macmillan.

Tobey, R. (1971). *The American ideology of national science: 1919–1930.* Pittsburgh: University of Pittsburgh Press.

Turner, F. J. (1920). *The frontier in American history.* New York: Henry Holt.

Turner, P. V. (1984). *Campus: An American planning tradition.* Cambridge, MA: MIT Press.

Tyler, R. W. (1950). *Basic principles of curriculum and instruction.* Chicago: University of Chicago Press.

Vassar, R. L. (1965). *Social history of American education* (2 Vols.). Chicago: Rand McNally.

Vincent, J. H. (1886). *The Chataugua movement.* Boston: Chataugua Press.

von Glasersfeld, E. (1992). Questions and answers about radical constructivism. In M. K. Pearsall (Ed.), *Scope, sequence, and coordination of secondary school science, Vol. II: Relevant research* (pp. 169–183). Washington, DC: National Science Teachers Association.

Wayland, F. (1842). *Thoughts on the present collegiate system in the United States.* Boston: Gould, Kendall, & Lincoln.

Weiss, I. (1987). *Report of the 1985–86 National Survey of Science and Mathematics.* Research Triangle Park, NC: Research Triangle Institute.

Welter, R. (1962). *Popular education and democratic thought in America.* New York: Columbia University Press.

Welter, R. (Ed.). (1971). *American writings on popular education: The nineteenth century.* Indianapolis: Bobbs-Merrill.

Whitty, G. (1985). *Sociology and school knowledge: Curriculum theory, research and politics.* London: Metheun.

Wilson, L. G. (Ed.). (1979). *Benjamin Silliman and his circle.* New York: Science History Publications.

Young, R. (1986). Introduction. In L. Levidow (Ed.), *Radical science essays* (pp. 1–14). London: Free Association Books.

Index

C

Child psychology, 217–218
  early childhood education in
    1960s, 222–223
  origins of, 113–118
  preschool education, 171
Civil War, 105
Cold War era
  academic competition in, 199–
    200
  Academism in, 189–191
  college enrollment in, 195
  curriculum developments in,
    194–195, 196–198
  influence of Conant in, 201–208
  liberal arts tradition in, 192–193
  National Science Foundation,
    208–215
  perception of science in, 193–
    194, 200–201
Columbia University, 24
Common school movement, 40,
  62–68, 136
Communist party, 162–163
Competency–based instruction,
  230
Conant, James Bryant, 201–208
Constructivism, 265–266
Cooper, James Fennimore, 59
Cooper Union, 81
Cornell University, 124
Counts, George, 186
Creationism, 242
Cremin, Lawrence, 12, 41–42
Cubberley, Ellwood P., 134
Curriculum. *See also* Science curri-
  culum
  Cold War era, 192–193, 194–
    195, 196–197, 201–208
  conservative trends in 1970s,
    228–230, 246
  development of theory in, 215–
    216
  early 20th century, 163–164
  Emerson's approach, 73–75

  experimental forms in 1960s/
    1970s, 227–228
  faculty involvement in design of,
    150
  Mann's design, 72
  National Science Foundation
    design, 210–215
  in 1980s, 254–260, 263–264,
    267–268
  Progressivism in design of, 144–
    148, 153–156, 165–168,
    183–187
  scientific approach to design of,
    115–119
  scientization of humanities, 152–
    153
  social concerns of 1960s/1970s
    and, 232–233
  Tyler's design, 215–218

D

Dartmouth, 24
Darwinian theory, 107, 109
Day, Jeremiah, 50–51
Developmental psychology, 217–
  218
  development of preschools, 171
  late 19th-century formulations,
    114–118
Dewey, John, 26, 110, 132, 135–
  144, 274–275
Discipline, Mann on, 71
Doctoral studies, 154

E

Edison, Thomas, 125
Education, role of
  Academist tradition, 21–24
  American attitudes toward, 3–8
  in Cold War era, 196
  continuity in American percep-
    tion of, 3–8, 273–276